Evolution by Neofunctionalization

The Role of Tumors in the Origin of New Cell
Types, Tissues and Organs

Evolution by Tumor Neofunctionalization

The Role of Tumors in the Origin of New Cell Types, Tissues and Organs

Andrei P. Kozlov

The Biomedical Center, St. Petersburg, Russia

AMSTERDAM • BOSTON • HEIDELBERG • LONDON
NEW YORK • OXFORD • PARIS • SAN DIEGO
SAN FRANCISCO • SINGAPORE • SYDNEY • TOKYO

Academic Press is an imprint of Elsevier

Academic Press is an imprint of Elsevier
The Boulevard, Langford Lane, Kidlington, Oxford, OX5 1GB, UK
225 Wyman Street, Waltham, MA 02451, USA

First published 2014

Notices
Knowledge and best practice in this field are constantly changing. As new research and
experience broaden our understanding, changes in research methods, professional practices,
or medical treatment may become necessary.

Practitioners and researchers must always rely on their own experience and knowledge in
evaluating and using any information, methods, compounds, or experiments described herein.
In using such information or methods they should be mindful of their own safety and the
safety of others, including parties for whom they have a professional responsibility.

To the fullest extent of the law, neither the Publisher nor the authors, contributors, or editors,
assume any liability for any injury and/or damage to persons or property as a matter of
products liability, negligence or otherwise, or from any use or operation of any methods,
products, instructions, or ideas contained in the material herein.

British Library Cataloguing-in-Publication Data
A catalogue record for this book is available from the British Library

Library of Congress Cataloging-in-Publication Data
A catalog record for this book is available from the Library of Congress

ISBN: 978-0-12-800165-3

For information on all Academic Press publications visit
our website at **store.elsevier.com**

This book has been manufactured using Print On Demand technology. Each copy is produced
to order and illustrations will appear in full colour in the electronic and printed version of
the book.

Working together
to grow libraries in
developing countries

www.elsevier.com • www.bookaid.org

To my wife Olen'ka

Contents

Acknowledgements

I wish to thank my wife Olga Kozlova for her continuous support without which this book would never have been finished. I thank my family, friends, colleagues and students for their patience to my virtual absence during the last three years devoted to the writing of this book.

I thank the team of the Biomedical Center, my collaborators and co-authors in many papers — first of all Ancha Baranova, Larisa Krukovskaya and Mark A. Zabezhinskiy for their contribution to experimental results supporting the hypothesis of evolution by tumor neofunctionalization; my younger colleagues Nilolai Samusik, Evgenii Shilov, Pavel Dobrynin, Ekaterina Matyunina and others for their contribution and enthusiasm; Boris Murashev for his input to experiments with fishes; Alexey Masharsky for his indispensible help in everyday laboratory life and in critical situations; and Tamara Kurbatova for her effort and her role in building the laboratory in which she was a first member many years ago. Special thanks are to Sergei Verevochkin for his permanent help in the lab and his technical assistance in drawing figures and diagrams; to Irina Kozodoi and Svetlana Gagarina for their help with many references and extensive correspondence and to Ludmila Vasil'eva, Olga Popova and Oxana Prohorchuk for their assistance.

I am grateful to my old friends Vladimir Evtushenko and Aleksander Emel'janov for many discussions, and to my long-term friends and oncological colleagues since pre-doctoral years, Vladimir N. Anisimov and Lev M. Bershtein, for their stimulating interest.

I wish to acknowledge the deep influence of the evolutionary school of St. Petersburg State University and the oncological schools of the N.N. Petrov Research Institute of Oncology and the National Cancer Institute, at different stages of my education and scientific career, on the origin and evolution of ideas presented in this book.

Introduction

This book is about the possibility of a positive role of tumors in the evolution of multicellular organisms. By pursuing such a paradoxical idea, I follow the maxim that "nothing in biology makes sense except in the light of evolution" [Dobzhansky, 1973]. Until now, the possibility of a positive evolutionary role of tumors was not addressed because different departments studied evolution and pathology. Interestingly, though, that the "evolution vs. pathology paradox" exists for mutations, and it may also exist for tumors, as I try to show in this book.

Tumors are widespread in multicellular organisms. Many of them are genetically or epigenetically determined and may be inherited. The majority (or at least a considerable part) of tumors may never kill their hosts. Tumors possess many features that could be used in evolution, and there are examples of tumors that have indeed played an evolutionary role. There is a lot of evidence on the convergence of tumorigenic and embryonic signaling pathways and on the connection of tumors with defects in differentiation, suggesting that tumors might have participated in the evolution of ontogenesis (evo-devo), specifically in the addition of the final stages of ontogenesis.

Multicellular organisms need extra cell masses for the expression of evolutionarily novel genes which originate in the DNA of germ cells, for the origin of new cell types and for building new organs and structures which constitute evolutionary innovations and morphological novelties. The source of cellular material used for the tremendous construction of body-plans in the evolution of Bilateria and Vertebrata is not known. It is generally assumed that the cellular material was somehow provided, as far as complexity has evolved.

I believe that tumor processes, in particular heritable tumors, provided evolving multicellular organisms with extra cell masses for the expression of evolutionarily novel genes, which originated in the germ plasm of evolving organisms, and for construction of morphological novelties. I formulate the hypothesis of evolution by tumor neofunctionalization, which I think is complementary to Susumu Ohno's hypothesis of evolution by gene duplication [Ohno, 1970], and present supporting evidence in this book.

Chapter 1

The Modern Synthesis of Evolutionary Biology and the Health Sciences

The synthesis of evolutionary biology and the health sciences is emerging. New disciplines – Darwinian medicine, evolutionary epidemiology and evolutionary oncology – attempt to apply the evolutionary approach to their corresponding traditional areas of research.

Darwinian medicine and evolutionary epidemiology are overlapping disciplines. P.W. Ewald [Ewald, 1994] suggests that Darwinian medicine should focus more on the individual patient, whereas evolutionary epidemiology focuses on the spread of diseases, i.e. the relationship is similar to that of medicine and epidemiology. At the same time, G.C. Williams and R.M. Nesse included both types of evolutionary considerations in their well-known paper "The Dawn of Darwinian Medicine" [Williams and Nesse, 1991].

From the standpoint of Darwinian medicine, many diseases and health conditions have evolutionary origins. For example, senescence may be a result of the selection of traits that are advantageous at the early ages but are associated with adverse effects later in life. Both nausea in pregnancy and allergy may be adaptations against toxins [Williams and Nesse, 1991].

Darwinian medicine attempts to address the general issues of evolutionary adaptedness. Human ancestors evolved in and adapted to the physical environment of the Pleistocene savannah [Orians, 1980; Williams and Nesse, 1991; Cerling et al., 2011]. Since then, the human environment has changed dramatically due to achievements of civilization, with elimination of many of the former factors of selection and creation of new ones. On the other side, from a genetic standpoint, humans are still Stone Age hunter-gatherers. This discordance does not affect reproductive success. Rather, it promotes chronic degenerative diseases that have their main clinical expression in the post-reproductive period. That is why many modern diseases, including obesity, diabetes mellitus, hypertension, atherosclerosis, dental caries, myopia, many cancers etc. may be called the "diseases of civilization" [Eaton et al., 1988], in accordance with the Darwinian medicine approach.

Diseases such as obesity, diabetes and high blood pressure may arise because our bodies are poorly adapted to the modern diet, which is rich in fat, sugar and salt. For millions of years, human ancestors evolved to eat a diet relatively high in protein and low in carbohydrates and fat. Only 10,000 years ago, when humans began to domesticate plants and animals, the big dietary shift brought a new Western-type diet — cereal grains, sugars, milk, refined fat and salt. During most of human evolution, our ancestors seldom ate these foods. Humans were well adapted for lean meat, fish, shellfish, insects and highly diverse plant foods, including fruits and root vegetables. Researchers think that humans need more time to become fully adapted to the modern diet. Currently, the epidemic of obesity is spreading throughout the world, especially into ethnic groups which until recently had remained more carnivorous [Gibbons, 2009; Lindeberg, 2009]. With the dietary shift came also an increase in cancer, which appears less frequently in hunter-gatherers and many traditional societies [Coffey, 2001; Cordain et al., 2005; Michels, 2005]. High caloric intake increases the risk of many cancers [Giovannucci, 2003; Hursting et al., 2003] and caloric restriction leads to a reduction in cancer rates [Hursting et al., 2003; Sell, 2003].

Evolutionary epidemiology assesses how traditional epidemiological characteristics such as lethality, illness, transmission rates, virulence and prevalence of infection change over time in the process of co-evolution of parasites and their hosts [Ewald, 1994]. Do parasites evolve toward benign coexistence with their hosts? Are symptoms adaptive? Which symptoms are the defense by the host and which represent a manipulation of the host? Should we treat the symptoms (e.g. should we use aspirin and other anti-inflammatory drugs during viral infections)? These are the kinds of questions that evolutionary epidemiology formulates and tries to answer [Williams and Nesse, 1991; Ewald, 1994]. The view that selection reduces virulence of the pathogen over time has been replaced by a more complex conception. It was established that pathogens with environmental reservoirs, transmission vectors or resistant spores have a higher level of virulence than directly transmitted pathogens. The lethality of vector-borne diseases is significantly greater than that of directly transmitted pathogens. Vector-borne transmission leads to relatively benign parasitism in the vector and severe parasitism of the vertebrate hosts. The virulence of diarrheal pathogens is positively associated with their tendencies for waterborne transmission. Attendant-borne transmission favors the more rapidly replicating and hence more virulent pathogens. Wartime conditions may also enhance the virulence of pathogens. Evolutionary epidemiology specifies that interventions should pursue not only the short-term benefit (i.e. the reduction in the disease transmission), but also the long-term benefit of evolutionary reduction in the parasite's virulence [Ewald, 1994; Dethlefsen et al., 2007].

Currently, attempts to apply the methods and concepts of evolutionary biology to studies of the different aspects of tumor growth are gaining popularity. They mainly deal with the somatic evolution of tumor cells and selection in populations of cells, rather than individuals [Boland and Goel, 2005; Merlo et al., 2006; Morange, 2012]. Competition between the individual cells within the single animal and selection of mutations that confer on a cell an increased survival advantage lead to cancer progression. Natural selection at the cellular level is harmful to the macro-organism [Cairns, 1975]. P.C. Nowell suggested a hypothesis of the clonal evolution of tumor cell populations [Nowell, 1976]. According to this hypothesis, "Tumor initiation occurs … by an induced change in a single previously normal cell which makes it 'neoplastic' and provides it with a selective growth advantage over adjacent normal cells. Neoplastic proliferation then proceeds, either immediately or after a latent period. From time to time, as a result of genetic instability in the expanding tumor population, mutant cells are produced. … Nearly all of these variants are eliminated, because of metabolic disadvantage or immunologic destruction…, but occasionally one has an additional selective advantage with respect to the original tumor cells as well as normal cells, and this mutant becomes the precursor of a new predominant subpopulation. Over time, there is sequential selection by an evolutionary process of sublines which are increasingly abnormal, both genetically and biologically. … Ultimately, the fully developed malignancy as it appears clinically has a unique, aneuploid karyotype associated with aberrant metabolic behavior and specific antigenic properties, and it also has the capability of continued variation as long as the tumor persists" [Nowell, 1976].

Important results have been obtained by studies of somatic evolution of tumor cells, including understanding of the development of tumor cell resistance to anti-cancer therapy, but this approach has nothing to do with the evolution of organisms.

This book is about the evolution of organisms with tumors and the role of tumors in the evolution of organisms, i.e. what I think evolutionary oncology should be about. The author has been working in this direction since the late 1970s [Kozlov, 1979]. My early papers on evolutionary oncology [Kozlov, 1979, 1983, 1987, 1988, 1996] approximately coincided with the appearance of the first publications on Darwinian medicine and evolutionary epidemiology. It means that during this period of the twentieth century, different branches of health science became mature for evolutionary generalizations, although comparative oncology started much earlier (see Chapter 3).

Chapter 2

Evolution and Pathology

Darwinian medicine and evolutionary epidemiology are looking for advantages which evolutionary biology could provide to the health sciences. On the other side, evolutionary biologists would be interested in elucidation of what role different pathologies could play in evolution. There were few studies of the latter kind, though, which is explained by the division of interest between medicine and biology, and by the fact that different people study pathologies and evolution. However, there were several deep observations and propositions on this issue. One of them is the notion of "the hopeful monsters." This term was introduced by E. Bonavia and R. Goldschmidt to express the idea that mutants producing monstrosities may play a role in macroevolution. Both authors suggested that monstrosities may cause significant adaptations, permit the occupation of new environmental niche and produce new types of organisms in a single large step [Bonavia, 1895; Goldschmidt, 1940]. The examples of monstrosities that Richard Goldschmidt gives in his major book, *The Material Basis of Evolution*, include mutants reducing the extremities which occur in man, in mammals and in birds; hairlessness and taillessness mutations in mammals; bulldog-head mutation in vertebrates from fishes to mammals; wing rudimentation in many groups of insects and birds; reduced eyes in insects, crustaceans and mammals; telescope eyes in fishes; and many others. Gouldschmidt points out that these monstrosities are considered as taxonomic traits and as adaptations to special environmental conditions. He concludes the corresponding chapter by assertion that "the hopeful monster is one of the means of macroevolution by single large steps" [Goldschmidt, 1940].

2.1 PATHOGENS AND PATHOLOGIES MAY HAVE ADAPTIVE AND/OR EVOLUTIONARY IMPORTANCE

The psychological obstacle to the recognition of a positive role of tumors in evolution is the fact that malignant tumors are pathological. The prevailing view is that pathological processes cannot play a positive role in evolution.

However, there do exist examples of pathogens and pathologies having an adaptive and/or positive evolutionary significance.

Evolution by Tumor Neofunctionalization.

Viruses may play an evolutionary role by transferring genes between different groups of organisms [Anderson, 1970; Reanney, 1974; Zdanov and Tikhonenko, 1974]. For example, polydnaviruses resemble the recombinant viral vectors used in gene therapy experiments and could be viewed as natural gene-delivery vehicles [Stoltz and Whitefield, 2009]. Recent studies in marine virology have shown that viruses move genetic material not only from one organism to another, but from one ecosystem to another. These studies have also shown that viral functional diversity, and its potential use for host adaptation and evolution, has been underestimated [Rohwer and Thurber, 2009]. Entomology provides examples of the mutualistic symbiosis among insects, bacteria, and viruses, in which viruses control the abundance of bacterial symbionts [Moran et al., 2005; Bordenstein et al., 2006]. It is suggested to start thinking about virus–host relationships in much broader terms, so as to include not only mutualism but also obligatory mutualism as exemplified by wasp-nudivirus symbiosis [Stoltz and Whitefield, 2009].

Like viruses, bacteria are generally known as a cause of infectious diseases. But many bacteria live as human symbionts. In the course of human and its symbiotic bacteria co-evolution mutualistic interactions important to human health developed. The examples of functional contributions of human gut bacterial symbionts include harvesting otherwise inaccessible nutrients and/or sources of energy from the diet, synthesis of vitamins, metabolism of xenobiotics, interacting with the immune system of the host, inhibition of host pathogens, detoxifying compounds harmful to the host, etc. Genomic studies have shown that the number of human symbiotic bacteria is greater than was previously anticipated. The human microbiome project is currently underway to fully understand the diversity of our microbial symbionts and their impact to human physiology [Dethlefsen et al., 2007; Turnbaugh et al., 2007].

For my consideration it is important that genetic or environmental changes can make symbionts pathogenic to the host, resulting in invasion of the host tissues and host immune response to clear away the infection. Similarly, tumors could be both symbiotic and pathological to their hosts.

The sickle cell trait provides some protection against *Plasmodium falciparum*, the parasite responsible for malaria [Allison, 1961; Livingstone, 1964]. It is known that sickle cell anemia is caused by an A-T transition in the hemoglobin A gene, which results in the synthesis of the alternative form of hemoglobin A called HbS. In regions with high malaria prevalence, the high frequency of the hemoglobin S allele is connected with the relative resistance of heterozygous carriers to malaria. Thus, the pleiotropic effect of the HbS allele leads to its preservation due to the positive selection of heterozygous carriers and provides an example of the positive adaptive significance of molecular pathology.

In a similar way, positively selected G6PD-Mahidol[487A] mutation – a common G6PD deficiency variant in Southeast Asia – reduces *Plasmodium vivax* density in Southeast Asians. Glucose-6-phosphate dehydrogenase

(G6PD) deficiency — the most common known enzymopathy — is associated with neonatal jaundice and hemolytic anemia usually after exposure to certain infections, foods, or medications. Strong and recent positive selection has targeted the Mahidol variant over the past 1500 years. The G6PD-Mahidol^{487A} variant reduces vivax, but not falciparum, parasite density in humans, which indicates that *Plasmodium vivax* has been a driving force behind the strong selective advantage conferred by this mutation [Louicharoen et al., 2009].

2.2 EVOLUTION VS. PATHOLOGY PARADOX OF MUTATIONS

The mutational process in general is the most dramatic example of a pathological process playing an important role in evolution. According to Michael Lynch and Bruce Walsh, "...the vast majority of new mutations are deleterious" [Lynch and Walsh, 1998, p. 352]. Multicellular organisms experience increased deleterious mutation rates [Lynch, 2007]. Other authors came to similar conclusions, although with varying estimates of degrees of mutation harmfulness [Eyre-Walker and Keightley, 1999; Eyre-Walker et al., 2002; Eyre-Walker et al., 2006; Kryukov et al., 2007]. Adam Eyre-Walker and co-authors came to the conclusion that a large majority (>70%) of amino acid mutations are strongly deleterious in all the species they investigated [Eyre-Walker et al., 2002]. More than 77% of amino acid alterations in hominid genes are deleterious [Mikkelsen et al., 2005].

Nevertheless, evolution is impossible without mutations. Mutation is one of four fundamental forces which govern evolution, together with natural selection, recombination and genetic drift.

The mutational process has two sides. On one hand, it provides new genetic material for selection and acts as a driving force of evolution. On the other hand, it disturbs balanced molecular mechanisms and thus generates various molecular diseases.

Psychologically, we accept this dichotomy implying that there are "good" and "bad" mutations. But the process that generates "good" and "bad" mutations is the same spontaneous process of genetic variation, which has different outcomes for different organisms and in evolutionary perspective.

Chapter 3

The Widespread Occurrence of Tumors in Multicellular Organisms

3.1 COMPARATIVE ONCOLOGICAL DATA ON THE PREVALENCE OF TUMORS IN DIFFERENT GROUPS OF MULTICELLULAR ORGANISMS

The origins of comparative oncology can be traced to 1802, when one of the scientific societies of Edinburgh raised the question of whether diseases reminiscent of human cancer occur in "brute creatures" [Dawe, 1969]. The cellular composition of human tumors was established in 1838 [Muller, 1838]. Before the end of the 19th century, tumors were discovered (with microscopic identification) in domestic animals [Leblanc, 1858], fishes [Bugnion, 1875], and mollusks [Ryder, 1887; Williams, 1890; Collinge, 1891].

In the 20th century, considerable efforts have been devoted to comparative oncological studies. Hundreds of papers on this topic were published, although some groups of multicellular organisms were studied less thoroughly than others. The first reviews of the field appeared [Willis, 1953; Huxley, 1958]. Special symposiums were organized, such as the Symposium on Neoplasms and Related Disorders of Invertebrate and Lower Vertebrate Animals held at the Smithsonian Institution, Washington, D.C. in 1968; several voluminous US National Cancer Institute monographs [Dawe and Harshbarger, 1969; Ziegler, 1980] and dozens of other monographs on comparative neoplasia were published; a special Registry of Tumors in Lower Animals was established by the U.S. National Cancer Institute to facilitate the comparative study of tumorigenesis and related disorders in invertebrate and poikilothermic vertebrate animals and to serve as a center of information and specimen reference material [Harshbarger, 1969]; a similar effort − The Veterinary Medical Data Program − was organized by the National Cancer Institute to collect data on the spontaneous occurrence of neoplasms in domestic animals [Priester, 1980]; etc.

All these efforts resulted in significant progress in the field of comparative oncology. Besides Vertebrata, tumors and tumor-like conditions have been described in Cephalochordata (lancelets), Urochordata (tunicates) and Echinodermata (echinoderms); in Protozoa, Coelenterata, Platyhelminthes, Annelida, Sipunculida, Arthropoda (predominantly in Insects) and Mollusca [Dawe and Harshbarger, 1969]. In a later publication, John Harshbarger, who for a long period of time directed the Registry of Tumors in Lower Animals, defined more exactly that "neoplasms have been found in all classes of cold-blooded animals in phylum Chordata (reptiles, amphibians, bony fish, cartilaginous fish, lampreys and hagfish) and in four of 28 phyla of invertebrates (mollusks, arthropods, platyhelminthes and cnidarians)... Additional unconfirmed neoplasm-like lesions have been reported or are being studied in several other invertebrate phyla including Sipuncula, Annelida and Echinodermata." Therefore, concludes Harshbarger, the number of phyla with characterized neoplasms is subject to expansion [Harshbarger, 1997]. Still later, Harshbarger stated that there is conclusive evidence that cancer exists throughout the phylogenetic tree [ZoomInfo, 2013].

As important steps, the discovery of neoplasm-like lesion in urochordates as early as in 1901 [Siedlecki, 1901] and the chemical induction of lethal atypical mesenchymal proliferation in echinoderms [Tchakhotine, 1938] should be mentioned. Tumors have even been discovered in dinosaur fossils [Moodie, 1917].

It was found that in plants, a variety of environmental impacts may result in continuing proliferative responses leading to tumor-like growths. Plant tumors may be caused by different pathogens or may be genetically determined [Dodueva et al., 2007]. Moreover, some plant tumors, including the so-called crown galls, which develop under the effect of the tumor-inducing bacterium *Agrobacterium tumefaciens* discovered in 1907 [Smith and Townsend, 1907], show many similarities with the tumors of animals.

The discovery and studies of transmissible tumors and their causative agents in new species or classes of animals and plants had exceptional value. For example, the discovery of transmissible avian sarcoma and its filterable causative agent in 1910–1913 [Rous, 1910, 1911; Rous and Murphy, 1913] significantly influenced the whole field of fundamental oncology and paved the way for the major breakthrough in molecular oncology in subsequent years. This discovery was honored by Nobel Prize awarded to Peyton Rous in 1967.

A lot of discussions were devoted to the difficulties in discrimination between neoplasia, hyperplasia, and response to injury or parasitic invasion in invertebrates and lower vertebrates. Many researchers questioned the occurrence of "true" neoplasms in these animals.

"...The paucity of information about neoplasms of mollusks and other invertebrates makes the distinction between neoplasia, hyperplasia, and response to injury or

infection difficult to determine confidently, mainly because there are no trenchant established criteria to diagnose neoplasia in these animals... In some cases gross lesions observed were thought to be neoplastic until histologic examination revealed them to be inflammatory responses or parasitic infestations. Growths do exist that are difficult to classify as either reactive or neoplastic" [Pauley, 1969].

"...In the invertebrates, ... many structures described as tumors in the literature are clearly hyperplasias or unusual proliferations of typical cellular components in response to injury or parasitic invasion and would not be considered tumors by the concepts developed in pathology of vertebrates. Points of difference that permit distinction between neoplastic and non-neoplastic changes are perhaps debatable and require experimental study not usually possible with the small numbers of tumors recognized among invertebrates in nature" [Sparks, 1969].

"It is recognized that a biological definition of neoplasia is not generally applicable ... because the biological potential of most tumors in these animals remains unknown. It is therefore necessary to base diagnostic judgments largely on the classical morphological criteria of the pathologist. Historically derived from studies of mammalian tissues, these criteria include increased mitotic rate, anaplasia, invasion, and metastasis. Their general validity among lower vertebrates and invertebrates is not yet firmly established" [Wellings, 1969].

Not only histological methodology of neoplastic versus non-neoplastic differentiation may cause the problem, but the line between hyperplasia and neoplasia could be as uncertain biochemically and physiologically as it is histologically [Grizzle and Goodwin, 1998]. This and similar opinions would suggest the possibility of the existence of some borderline tumor-like processes, a possibility that will be important for my future considerations of what processes could supply evolving multicellular organisms with extra cell masses for the origin of new cell types, tissues and organs.

Understanding of neoplasms achieved through studies of mammalian tumors and tumors of other higher vertebrates may not be applicable to tumors of lower vertebrates and invertebrates because of the biological differences[1].

---

1. Even in mammals, there are significant interspecies differences in tumor biology. Essential differences in cancer development between humans and rodents include tumor origin — commonly mesodermal sarcomas in mice compared with epithelial carcinomas in humans; carcinogenic risk factors — many rodent carcinogens are non-carcinogenic in humans and vice versa; the spectrum of common spontaneous tumors — there are no rodent strains with high incidence of spontaneous stomach, colon or bladder tumor that are common in humans; and spontaneous regression of tumors which occurs in infants but is rare in adult humans, whereas it is common in adult mice [Anisimov et al., 2005].

"...The differences between neoplasms of invertebrates and of vertebrates may be so great that we fail to recognize their nature and kinship" [Dawe, 1969].

"...The criteria used in mammalian classification schemes may not be adequate for lower vertebrates due to the biologic differences among vertebrates, and because these criteria have generally been developed in a clinical context to provide prognostic information that generally does not apply to lower vertebrates including fishes"[Groff, 2004].

For example, information concerning the clinical consequence or outcome of the majority of neoplasms in fishes is, on the whole, non-existent. The adverse effects of fish tumors are not always obvious and it seems that in wild fish neoplasms do not affect the size of fish populations. Neoplasms in fishes are generally less aggressive and more differentiated than neoplasms in mammals. They are mostly discrete, focal, benign neoplasms. The most important feature of few malignant fish neoplasms is local invasion: they rarely exhibit metastatic behavior. That is why invasiveness is the criterion used to determine malignancy of fish neoplasms.

The reasons for relative absence of metastatic behavior of fish neoplasms, although not completely elucidated, may include higher differentiation of fish tumor cells, differences in fishes' lymphatic systems and other anatomic differences from mammals, and lower temperatures in fishes compared with mammals. While in mammals a malignant neoplasm is generally associated with a poor prognosis, this may not necessarily be correct in fishes [Hayes and Ferguson, 1989; Machotka et al., 1989; Grizzle and Goodwin, 1998; Groff, 2004].

Seasonal tumors with complete regression have been described in fish [Getchell et al., 1998; Holzschu et al., 2003], a phenomenon not known in mammals. Several such lesions are described in different fish species, all tentatively attributed to retroviruses based on the observation of retrovirus-like particles and reverse transcriptase activity. Seasonal fish tumors develop and regress annually; for example, walleye dermal sarcomas (WDS) first can be seen in the fall and completely regress in the spring. WDS is a benign tumor which does not lead to the death of feral walleye, but when experimentally transmitted to walleye fingerlings WDS generates invasive tumors [Holzschu et al., 2003]. Another similar example is cutaneous hyperplasia caused by *Herpesvirus cyprini* in cyprinids, known as carp pox or fish pox. This is usually a benign, chronic epidermal disease which is more common during cooler seasons and spontaneously regresses after several months with the elevation of temperatures [Groff, 2004]. Basing on these and other data some authors argue that fish tumors show intermediate forms between malignancy and normalcy [Martineau and Ferguson, 2006].

A special term — "pseudoneoplasms" — was coined to designate non-neoplastic lesions that resemble neoplasms in ectothermic animals

[Harshbarger, 1984], but it probably does not exhaust the whole problem of hyperplasia versus neoplasia differentiation and the issue of borderline tumor-like processes. Categories of pseudoneoplasms would include virally induced hyperplasia or hypertrophy, some parasitic diseases, dysmorphogenesis (teratoid anomalies), unusual normal conditions (giant islets of endocrine pancreas in the liver or atypical sites of hematopoietic tissue), reactive lesions (metaplasia, regeneration, inflammation) and non-parasitic hyperplasia (goiter, ectopic thyroid, erythroblastic proliferation suggestive of pernicious anemia, and adenofibrosis). Thyroid hyperplasia in fishes represents an interesting example of a borderline tumor-like process. Because the teleost thyroid lacks a capsule and has a diffuse nature, hyperplastic thyroid tissue in fish is commonly characterized by invasiveness and metastasis, making it difficult to distinguish the lesions as either cancer or goiter [Harshbarger, 1984; Hoover, 1984; Grizzle and Goodwin, 1998].

There are, however, examples of malignant neoplasms in lower vertebrates and invertebrates, which may be considered "true" malignant neoplasms even from the point of view of mammalian classification. For instance, hepatobiliary neoplasms of rainbow trout exposed to aflatoxin are characterized by aggressive local invasion and metastasis (reviewed in [Groff, 2004]).

In mollusks, tumors originally were described as large, readily visible masses growing externally on the surface of the animal. Invasion of normal tissues by atypical cell types, abnormal arrangement of cells and mitotic figures within a suspect growth were observed. Uncontrolled growth with no apparent stimulus was suggested as one criterion of neoplasia in mollusks [Pauley, 1969].

But during the last several decades two malignant neoplasms of marine bivalves have been described − disseminated neoplasia and gonadal neoplasia. Disseminated neoplasia is characterized by the presence of large hemocytes of unknown origin. Gonadal neoplasia consists of a proliferation of undifferentiated germ cells. Both neoplasias fit the criteria of malignant tumors, including atypical structure with pleomorphic, undifferentiated cells; invasive growth; mitotic figures; metastasis; and progressive growth. Disseminated neoplasia can result in significant mortality of affected populations. Prevalence exceeding 90% has been reported in *Mya arenaria*. Gonadal neoplasia progresses slowly and mortality rates are low. This disease primarily affects *Mya arenaria* and *Mercenaria* spp. at prevalences up to 50%. The finding that prevalence of gonadal neoplasia is higher in hybrid *Mercenaria* spp. suggests a genetic etiology [Sparks, 1985; Barber, 2004].

The prevalence of neoplasms is not evenly distributed in different groups of multicellular organisms. Among invertebrates tumors are most widespread in insects and mollusks[2]. In vertebrates, tumors are most widespread in mammals [Fox, 1912; Ratcliffe, 1933; Lombard and Witte, 1959; Effron et al., 1977]. In lower vertebrates, the prevalence of tumors is highest in bony fishes where most types of tumors characteristic to mammals are encountered. In bony fishes, neoplasms have species or family specificities. For example, epidermal papillomas are common in flatfish, lymphomas and lymphosarcomas occur frequently in the pike, pseudobranchial tumors are found in gadoid fishes and nerve sheath tumors are found in snappers [Sinderman, 1990]. In Chondrichthyes the prevalence of tumors is much lower than in Osteichthyes. In chondrichthyan elasmobranches (sharks, skates and rays) and holocephalans (chimaeras) and other primitive fishes including African lungfish (*Protopterus spp.*; Family Protopteridae), bowfin (Family Amiidae) and the chondrostean fishes (sturgeons and paddlefishes; Order Acipenseriformes) neoplasms are rare but have been reported (reviewed in [Groff, 2004]). The prevalence of neoplasms continues to decrease downward through the cyclostomes, the cephalochordates, the hemichordates, and into tunicates [Dawe, 1969; Wellings, 1969][3,4].

Basing on these and other data discussed above, comparative oncology formulated the generalization that neoplasia could be a property of all or most living systems but tumors are more frequent among the "higher" forms, e.g. insects and vertebrates [Wellings, 1969] and in the evolutionarily more successful groups of organisms, e.g. in teleost fishes compared with cartilaginous fishes [Dawe, 1973][5]. The examination of more than 10,000 dinosaur vertebrae using fluoroscopy revealed tumors in hadrosaurs of the Cretaceous period. In these studies, tumors were not found in other dinosaur species, suggesting a species-specific predisposition to neoplasia in dinosaurs [Rothschild et al., 2003]. Hadrosaurs are one of the most successful groups among dinosaurs because this family (Hadrosauridae) is the most numerous and contains the greatest number of species (more than 40) as compared to the other dinosaur groups. Hadrosaur tumor prevalence data

---

2. Mollusks and insects were more frequently studied by biologists and their tumors more readily observed [Sparks, 1969].
3. The failure to recognize spontaneous neoplasms among invertebrates and lower vertebrates could be the result of relatively smaller effort [Wellings, 1969; Groff, 2004].
4. In invertebrates and cold-blooded vertebrates, "the largest number of spontaneous neoplasms is known in the bony fish — estimated to exceed 10,000 cases. Other groups..., i.e. rays and sharks, lampreys, hagfish, coral and flatworm, are estimated at approximately 100 or less. For example, there are 30—50 neoplasm cases in rays and sharks and less than ten in flatworms. The prevalence of known cases in a given group is influenced by many factors not the least of which is the number of scientists studying the group" [Harshbarger, 1997].
5. I connect the high prevalence of spontaneous neoplasms in bony fishes with the fish-specific genome duplication (3R) in the teleost lineage. As we will see below, novel genes which may originate as a consequence of genome duplication need extra cell masses for their expression.

thus support the connection between increased tumor susceptibility and evolutionary success.

3.2 ANCIENT ORIGIN AND CONSERVATISM OF CELLULAR ONCOGENES AND TUMOR SUPPRESSOR GENES

The first indication of widespread occurrence and conservatism of cellular oncogenes was demonstration that DNA sequences homologous to vertebrate protooncogenes are conserved in *Drosophila melanogaster* and *Caenorhabditis elegans* nematodes [Shilo and Weinberg, 1981].

Activated *ras* oncogenes have been described in fish and molluscan neoplasms [van Beneden, 1997][6]. The *myc* genes have been found in a wide variety of vertebrates, including mammals, birds, amphibians and fish [Atchley and Fitch, 1995]. The *myc* homologs with conserved basic functions have been found in *Hydra* [Hartl et al., 2010] and in sponge [Srivastava et al., 2010], i.e. they may be older than 600 million years. The important developmental regulator and oncogene beta-catenin was described in cnidarians, crustacean *Artemia salina*, insects, nematode *Caenorhabditis elegans*, annelids, echiurid *Urechis caupo*, sea urchins, ascidians, fish, frogs, chickens, mice, rats and humans [Schneider et al., 2003].

Wnt-1 was discovered as a protooncogene responsible for mammary tumors induced by the mouse mammary tumor virus [Nusse and Varmus, 1982]. Inappropriate activation of the Wnt signal transduction pathway plays a role in a variety of human and mice tumors [Nusse and Varmus, 1982; Nusse and Varmus, 1992; Peifer and Polakis, 2000; Polakis, 2000; Reya and Clevers, 2005; Fan and Eberhart, 2008]. The Wnt protein sequences have been conserved during a billion years of evolution, suggesting common mechanisms of action. For example, 11 of 12 vertebrate *Wnt* gene families were also described in cnidarians. Mutations in *Wnt* genes or Wnt pathway components lead to specific developmental defects [Nusse and Varmus, 1992; Parr et al., 1993; Cadigan and Nusse, 1997; Nusse, 2001; Kusserow et al., 2005].

Proteins homologous to tumor-suppressor protein p53, which inhibits tumor formation in mammals, and to its related proteins p63 and p73 have been isolated in a number of invertebrates [Schmale and Bamberger, 1997; Kelley et al., 2001; Cox et al., 2003; Muttray et al., 2005; Goodson et al., 2006; Muttray et al., 2007]. Although p53 was discovered first, p63 and p73 are evolutionarily older. p53 originated from the ancestral version of p63 and p73 more than 450 million years ago [Leslie, 2011]. Studies of the expression of homologs for human *p53* and *p73* genes in *Mya arenaria* suggest that they may serve similar functional roles to those in humans (Kelley et al., 2001). p53 mRNA and an N-terminal truncated isoform, deltaNp63/73 were

---

6. Gene symbols are italicized. Protein designations are the same as gene symbols but not italicized (http://www.gmb.org.br/nomenclature.html)

expressed during late-stage hemic neoplasia in the mussel *Mytilus trossulus* at levels higher than those seen in normal cells [Muttray et al., 2008]. Molluscan MDM homolog (MDM is a negative regulator of mammalian p53 family proteins) interacts with molluscan p53 and the *mdm* gene is differentially expressed together with *p53* and *ras* in neoplastic *M. trossulus* hemocytes [Muttray et al., 2010].

Oncogenes and tumor suppressor genes are characterized by higher intensities of purifying selection than other genes. In this respect, they are similar to "essential genes" identified by knockout experiments [Thomas et al., 2003]. But they may also participate in the origin of new genes. Studies of the structural and functional evolution of beta-catenin genes suggested that beta-catenin genes fulfilled adhesion and signaling functions in the last common ancestor of metazoans some 700 million years ago. A polarized epithelium organized by β- and α-catenin in the non-metazoan *Dictyostelium discoideum* predates metazoan origins. The role of catenins in cell polarity precedes the evolution of Wnt signaling and classical cadherins [Dickinson et al., 2011]. Gene duplications facilitated the evolution of beta-catenins with novel functions and allowed the evolution of multiple, single-function proteins (cell adhesion or signaling) from the ancestral, dual-function protein [Schneider et al., 2003].

It was shown that a considerable number of protein domains connected with cancer predates the origin of multicellularity, and the other group of cancer protein domains arose at about the same time as metazoans appeared [Domazet-Loso and Tauz, 2010a].

The sequencing of the genome of the sponge *Amphimedon queenslandica* has shown that it contains many genes implicated in cancer — tumor suppressor genes, oncogenes and genes associated with programmed cell death [Srivastava et al., 2010]. The presence of such genes indicates that the mechanisms which control tumor development are as old as multicellularity itself.

These and other results of molecular biologic and genomic studies of primitive multicellular organisms and invertebrates not only support the concept that tumors are characteristic to all multicellular organisms. Widespread occurrence and conservatism of cellular oncogenes and tumor suppressor genes also suggest that they have important physiologic and evolutionary roles in multicellular organisms.

As one of the discoverers of cellular oncogenes, J.M. Bishop, put it soon after B.Z. Shilo and R.A. Weinberg's discovery, "the logic of evolution would not permit the survival of solely noxious genes" [Bishop, 1983].

3.3 THE WIDESPREAD OCCURRENCE OF TUMORS SUGGESTS THAT THEY MAY BE EVOLUTIONARILY MEANINGFUL

The wide occurrence of tumors and tumor-like processes in multicellular organisms, tumors' connection to evolutionary success and progressive evolution and the wide distribution and conservatism of cellular oncogenes suggest that tumors and/or some tumor-like processes could play a role in the evolution of multicellular organisms. Relatively lower malignancy of tumors in low forms of multicellular organisms supports such a possibility. There are features of tumors that could be used in evolution: tumors generate extra cell masses which are functionally not necessary for the organism and could be used as a building material for new cell types, tissues and organs; many genes usually inactive in normal cells are activated in tumor cells; tumor cells can differentiate with concomitant loss of malignancy; and tumors may participate in morphogenetic processes.

Evolution would not miss the opportunity to use extra cell masses characterized by high biosynthetic and morphogenetic potential for the creation of new forms of multicellular organisms. Tumors could represent those underlying background proliferative processes that eventually led to realization of the tendency towards the increase in the number of cells and cell types in multicellular organisms.

Chapter 4

Features of Tumors that Could Be Used in Evolution

4.1 UNUSUAL GENES AND GENE SETS ARE ACTIVATED IN TUMORS AND MAY PARTICIPATE IN THE ORIGIN OF NEW CELL TYPES

It has long been known that many tumors and cell lines derived from tumors are able to produce proteins which are not characteristic of the tissues and cell types of their origin. This phenomenon has been termed "ectopic syntheses." An example is polypeptide hormone production by neoplasms of non-endocrine tissues [Omenn, 1970; Frohman, 1991]. The spectrum of ectopic hormones and the variety of hormone-producing tumors is impressive and includes dozens of different hormones and non-endocrine tumors [Gurchot, 1975; Frohman, 1991]. Activation of related structural genes is the cause of this phenomenon [Omenn, 1970]. Adrenocorticotropic hormone (ACTH) is the most frequently observed ectopic hormone produced by neoplasms [Torosian, 1988]. The ectopic production of ACTH was first described in 1928 in a patient with small-cell carcinoma of the lung [Brown, 1928]. ACTH has been found in cancer of the thyroid [Goldberg and McNeil, 1967], parathyroid [Friedman et al., 1966], thymus [Miura et al., 1966], lung [Marks et al., 1963], ovary [Nichols et al., 1962], testicle [Schoen et al., 1961], breast, prostate and pancreas [Liddle et al., 1965; Dollinger et al., 1967]; in chorionepithelioma of the liver [Piper, 1960]; and in carcinoid tumors [Steel et al., 1967]. Human chorionic gonadotropin (HCG) has been found in carcinoma of the lung [Fusco and Rosen, 1966; Rosen et al., 1968] and other cancers [Braunstein et al., 1973]. The HCG-β subunit gene is expressed in cancer cells of different types and origins [Acevedo et al., 1995]. Human chorionic thyroid-stimulating hormone (HCT), besides chorion [Hennen, 1965, 1966a], was found in bronchial carcinoma [Hennen, 1966b] and in male [Hennen, 1966c] and female [Odell et al., 1967] choriocarcinomas. A tumor may produce more than one hormone activity, for example, both ACTH and melanocyte-stimulating hormone (MSH) [Omenn, 1970]. Ectopic production of multiple hormones (ACTH, MSH and gastrin) by a single malignant islet-cell tumor of the pancreas has been reported [Law et al., 1965].

The term "tumor cell plasticity" is currently used and implies the loss of specific gene expression and the concomitant aberrant expression of genes normally active in other cells. Such plasticity is especially pronounced in malignant tumor cells, as exemplified by vasculogenic mimicry manifested as tumor cell expression of the genes that are normally active in endothelial cells. In the case of melanoma, the aggressive phenotype can engage in vasculogenic mimicry by expressing endothelial-associated genes, such as vascular endothelial cadherin (VE-Cadherin). Vasculogenic mimicry has been demonstrated in several tumor types, including breast carcinoma, prostatic carcinoma, ovarian carcinoma, lung carcinoma, synoviosarcoma, rhabdomyosarcoma, phaeochromocytoma and Ewing's sarcoma [Hendrix et al., 2003, 2007].

A general phenomenon is expression of carcinoembryonic antigens in tumors. Originally, alpha-fetoprotein (AFP) was discovered by Garry I. Abelev and co-authors [Abelev et al., 1963a,b] to be associated with a mouse hepatoma and Y.S. Tatarinov discovered an analogous protein associated with human hepatomas [Tatarinov, 1964]. The carcinoembryonic antigen(s) (CEA) were found in human fetal gut and in human colonic carcinoma [Gold and Freeman, 1965] and in a wide variety of tumors of both endodermal and nonendodermal origin. Since then, the number of different embryonic and fetal antigens found to be associated with tumors has been rapidly increasing [Coggin and Anderson, 1974].

CEA is a member of the immunoglobulin supergene family. The CEA gene family is comprised of at least 17 members. Both transcriptional and post-transcriptional control mechanisms participate in variable expression of CEA. The degree of methylation of 5' of CEA gene is inversely correlated to CEA expression. The CEA gene family probably originated a number of times, in parallel, in different mammalian orders [Thomas et al., 1990].

Cancer/testis antigen genes (CTA or CT genes) code for a subgroup of tumor antigens expressed predominantly in testis and different tumors [Zendman et al., 2003a,b; Simpson et al., 2005; Chen et al., 2006; Caballero and Chen, 2009; Wu and Ruvkun, 2010]. To date, the CTDatabase (http://www.cta.lncc.br) includes 265 CT antigen genes and 149 CT antigen gene families. Among human tumors, CT antigens are expressed in melanoma, bladder cancer, lung cancer, breast cancer, prostate cancer, sarcoma, ovarian cancer, hepatocellular carcinoma, hematologic malignancies, etc. [Hofmann et al., 2008; Caballero and Chen, 2009; Cheng et al., 2011; Fratta et al., 2011]. These genes are more frequently expressed in high-grade late-stage cases of cancer [Simpson et al., 2005; Caballero and Chen, 2009]. The class of cancer/testis antigen genes is discussed in more detail in Part 11.1.4.1.

Differential expression of human endogenous retroviruses (HERVs) is described in many human tumors: HERV-K family — in teratocarcinoma

[Boller et al., 1993], in breast cancer [Wang-Johanning et al., 2001], in urothelial and renal cell carcinomas [Florl et al., 1999], and in stomach cancers [Stauffer et al., 2004]; HERV-E – in prostate carcinoma [Wang-Johanning et al., 2003]; HERV-H – in leukemia cell lines [Lindeskog and Blomberg, 1997] and in cancers of the small intestine, bone marrow, bladder and cervix [Stauffer et al., 2004]. Although originally it was thought that HERVs are transcriptionally silent in most normal tissues [Florl et al., 1999], *in silico* data suggest that HERV-derived RNAs are more widely expressed in normal tissues than originally anticipated [Stauffer et al., 2004].

Tumor antigens discovered in humans include oncofetal antigens, differentiation antigens, mutational antigens, overexpressed cellular antigens, viral antigens, and cancer/testis antigens [Schreiber et al., 2011]. The National Cancer Institute's effort to prioritize cancer vaccine target antigens resulted in a list of 75 cancer antigens, many of which have absolute tumor specificity. Contrary to tumor serum markers, cancer vaccine target antigens preferably are expressed on the cell surface with no or little circulating antigen [Cheever et al., 2009].

Numerous molecular markers of human malignant lesions have been described for detecting and localizing tumors and for monitoring the response of tumor to therapy (reviewed in [Torosian, 1988; Pandha and Waxman, 1995; Cairns and Sidransky, 1999; Chatterjee and Zetter, 2005; Manne et al., 2005]). The first tumor marker was identified by Bence-Jones [Bence-Jones, 1847]. The "Bence-Jones protein" is circulating monoclonal immunoglobulin found in myeloma, Waldenstrom's macroglobulinemia and non-Hodgkin's lymphomas. It occurs in the serum or urine of almost a hundred per cent of patients [Pandha and Waxman, 1995].

Tumor markers were originally defined as substances present at high levels in serum (i.e. serum markers) of patients with cancer. M.H. Torosian included in his table of tumor markers oncofetal antigens (CEA and AFT), placental proteins (human chorionic gonadotropin, human placental lactogen, placental alkaline phosphatase, Regan isoenzyme), hormones (calcitonin, adrenocorticotropic hormone and antidiuretic hormone), enzymes (prostatic acid phosphatase, bone alkaline phosphatase), catecholamine metabolites, and "miscellaneous" (polyamines, acute phase proteins and immunoglobulins) [Torosian, 1988]. He intentionally did not discuss tumor-associated antigens of the cell surface and intracellular proteins specific to malignant cells.

Additional biomarkers developed in the 1980s were CA 19-9 for colorectal and pancreatic cancer, CA 15-3 for breast cancer and CA-125 for ovarian cancer. These are altered or abnormal carbohydrate structures commonly found on the tumor cell surface. CA 15-3 is a carbohydrate antigen and CA 19-9 is a Lewis blood-group glycolipid antigen. CA-125 is a 200 kDa carbohydrate antigen normally present during embryonic development. The function of the protein bearing the CA-125 epitope is unknown [Pandha and Waxman, 1995; Chatterjee and Zetter, 2005]. Addition of other markers associated with epithelial ovarian cancer (other glycoproteins, hepatic and acute phase proteins, cytokines, growth factors and serine proteases) to CA-125 was associated with higher sensitivity and detection rates than either marker alone [Husseinzadeh, 2011].

Prostate-specific antigen (PSA), a glycoprotein from ductal epithelial cells within the prostate, is the cancer biomarker used to detect disease at the early stage. The serum PSA test was widely used in screening for prostate cancer [Pandha and Waxman, 1995; Chatterjee and Zetter, 2005].

Tumor-associated antigens of the cell surface and intracellular proteins specific to malignant cells originally were not considered as potential tumor markers because of the difficulties connected with the access to biological material and lack of appropriate sensitivity. Currently, cellular (or tissue) biomarkers are represented among tumor markers, and they include not only protein based markers, but also RNA- and DNA-based markers [Cairns and Sidransky, 1999][1]. In their classification of tumor markers, Pandha and Waxman separate tumor-associated antigens that include viral antigens (polyoma, SV40), MHC-related antigens (H-2k antigen), enzymes (PAP, NSE and PLAP), oncogene products (c-myc, c-erbB2) and cytogenetic markers (Ph' chromosome) [Pandha and Waxman, 1995].

Cancer stem cells (CSCs), or tumor-propagating cells (TPCs), can be defined by the expression of the specific cell surface markers: the *ALDH* gene superfamily that encodes detoxifying enzymes; Bmi1 polycomb protein; CD24, a sialoglycoprotein that acts as a ligand for P-selectin; CD44, a transmembrane glycoprotein that binds hyaluronan; CD90, a glycosylphosphatidylinositol-anchored protein; CD105, a type I integral membrane protein; CD117, a tyrosine kinase receptor for SCF; CD133, a member of the prominin family of pentaspan membrane proteins; CD166, an activated leukocyte cell-adhesion molecule (ALCAM); EpCAM, a glycosylated type I integral membrane protein; ABCG2, a member of the ABC superfamily of membrane transporters; etc. Some cell surface markers (CD133, CD44, CD24 and THY1) are found in several cancers. For example, CD133 has been the best marker for brain CSCs. CSCs of colon, pancreas, lung, ovarian and endometrial tumors also express CD133. Most human breast cancers contain cells with the $CD44^+CD24^{-/low}$ phenotype. In hematological malignancies, the $CD34^+CD38^-$ signature identifies the majority of TPCs. CD166 combined with CD44 or CD24 combined with CD29 mark colon CSCs [Vermeulen et al., 2008; Alison et al., 2010; Hubbard and Gargett, 2010].

The list of tumor biomarkers officially registered for clinical use is not extensive [Chatterjee and Zetter, 2005l; Manne et al., 2005], but the number of potential tumor markers is considerable.

---

1. On the other hand, not only proteins but also DNA and RNA can be detected in the circulation [Cairns and Sidransky, 1999].

Molecular hybridization approaches (i.e. saturation hybridization with unique DNA probes) have demonstrated that tumors can express numerous genes that are inactive in corresponding normal cells (see, e.g. [Supovit and Rosen, 1981]). Moreover, saturation hybridization experiments with "combined" RNA preparations (i.e. mixtures of RNAs extracted from many different normal tissues and organs, and different stages of development) and molecular probes recirculated after hybridization with tumor RNA have shown that hundreds of medium-size genes may be expressed in tumors but not in any experimentally available normal cells [Evtushenko et al., 1989; Kozlov et al., 1992; Kozlov, 2008]. Similar results were later obtained by *in silico* comparison of millions of sequences (expressed sequence tags, EST) from thousands of normal and tumor cDNA libraries using the so-called computer-assisted differential display (CDD) approach and special software tools. It was shown that tumors express hundreds of sequences that are not expressed in normal tissues [Baranova et al., 2001]. The tumor specificity of many of these sequences was later experimentally confirmed [Krukovskaya et al., 2005].

The use of DNA microarrays to compare the global transcriptional profiles (or gene expression signatures) of cancer tissue and corresponding normal tissue also generates hundreds of genes differentially expressed in cancer relative to normal tissue [Liotta and Petricoin, 2000; Rhodes et al., 2004]. The use of these modern genome- and proteome-wide methodologies (DNA and protein arrays, computational high-throughput approaches, etc.) thus led to the discovery of hundreds of new tumor marker candidates [Sikaroodi et al., 2010]. A transcriptional profile that is commonly activated in various types of undifferentiated cancer was characterized [Rhodes et al., 2004]. In melanoma research, comparative global gene analyses of aggressive and poorly aggressive human melanoma cell lines have revealed that in aggressive tumors genes characteristic to various cell types are expressed with concomitant downregulation of genes specific to their parental melanotic lineage [Bittner et al., 2000; Hendrix et al., 2007].

The accumulating data suggest that the fraction of transcribed genome is greater than was earlier anticipated. Advanced saturation hybridization studies with mixtures of RNA mentioned above demonstrated 56% expression of unique DNA sequences in rats [Evtushenko et al., 1989]. This value was cited as the highest level of expression in mammals ever detected experimentally [Lennon and Lehrach, 1991]. Still more expression was demonstrated in tumors [Kozlov et al., 1992; Kozlov, 2008]. Genome-wide transcription studies performed within the ENCODE project have shown that 75% of the human genome is transcribed collectively in the 15 cell lines studied [Djebali et al., 2012]. Importantly, RNA samples that were used in genome-wide expression studies were obtained from tumor or immortalized cell lines [Maher, 2012].

A large portion of noncoding regions in the human genome has been shown to be transcribed by means of different approaches [Baranova et al., 2001; Kapranov et al., 2007]. The corresponding transcripts are called noncoding RNAs (ncRNAs). The ENCODE consortium proposed to term such putative noncoding molecules the "transcripts of unknown function" (TUFs) [Kapranov et al., 2007]. The biological significance of this phenomenon is unclear. Of more than 200 tumor-specific sequences described by *in silico* analysis [Baranova et al., 2001], about half are represented by ncRNAs. We suggest that noncoding sequence expression in tumors is associated with the origin of novel genes and gene networks and the new functions [Kozlov, 1996, 2010; Baranova et al., 2001; Krukovskaya et al., 2005; Kozlov et al., 2006].

From all the examples mentioned above we may conclude that many unusual genes, gene sets and noncoding sequences may be activated in tumor cells.

The activation of transcription in tumors is associated with epigenetic changes, i.e. changes in DNA methylation, histone modifications and changes in chromatin structure. All examined benign and malignant tumors demonstrated global reduction of DNA methylation [Feinberg and Vogelstein, 1983; Gama-Sosa et al., 1983; Goelz et al., 1985; Schulz, 1998; Baylin et al., 1998; Lin et al., 2001; Robertson, 2001; Ehrlich, 2002, 2009; Jones and Baylin, 2002, 2007; Kisseljova and Kisseljov, 2005; Zhu, 2006; Cheung et al., 2009; Pogribny and Beland, 2009; Schulz and Hoffmann, 2009][2]. DNA hypomethylation may play a causal role in tumor formation, as shown by induction of tumors by genomic hypomethylation in DNA methyltransferase 1 (*Dnmt1*) allele hypomorphic mice [Gaudet et al., 2003]. Global loss of acetylation at Lys16 and trimethylation at Lys20 is a common hallmark of human cancer [Fraga et al., 2005]. Changes in DNA methylation and chromatin remodeling events are loosely specific and affect large portions of the genome. They may result not only in transcriptional activation of genes related to tumor progression, but also in non-specific transcriptional activation of sequences located in the same genome areas. Some of these sequences, e.g. endoretroviruses [Lower et al., 1993; Sauter et al., 1995; Anderson et al., 1998], intragenomic endoparasitic DNA such as L1 [Schulz, 2006] and Alu [Berdasco and Esteller, 2010], and processed pseudogenes [Moreau-Aubry et al., 2000; Berger et al., 2005; Zheng et al., 2005], are normally not expressed in the differentiated cells.

The fact that many unusual sequences that are not expressed in normal cells are transcribed in tumor cells illustrates the magnitude of the biosynthetic potential of tumor cells that may be used in evolution. Novel genes and gene sets could be expressed in tumor cells and novel gene networks could be formed to participate in the origin of novel cell types, tissues and organs.

---

2. At the same time, tumor-type-specific CpG island hypermethylation in the promoter regions of tumor suppressor genes is a general characteristic of cancer cells [Baylin et al., 1998; Berdasco and Esteller, 2010].

4.2 TUMOR CELLS CAN DIFFERENTIATE WITH THE LOSS OF MALIGNANCY THAT MAY LEAD TO THE ORIGIN OF NEW CELL TYPES

Another tumor cell property which might be used in evolution of organisms is the ability to differentiate with the loss of malignancy.

Tumor cells are characterized by deviation from typical differentiation, i.e. metaplasia, dysplasia and anaplasia. It reflects the tumor's origin from undifferentiated stem cells, as currently believed. Nevertheless, in many malignant tumors some cells exhibit the signs of aberrant differentiation that does not lead to normal histogenesis or organogenesis. For example, malignant neuroblastoma cells in children often spontaneously increase differentiation, as a result of which the tumor matures into benign ganglioneuroma that completely lose malignancy [Evans et al., 1976].

Tumors of terminally differentiated cells are not described. When normal cells differentiate, they usually lose their ability to continue cell division. In a similar way, tumor malignancy decreases with differentiation.

The ability to differentiate both spontaneously and under the influence of various differentiation-inducing factors is especially pronounced in teratocarcinomas. The tissues derived due to differentiation of the embryonal carcinoma cells are mostly benign [Pierce and Dixon, 1959; Pierce et al., 1960; Pierce and Verney, 1961]. Normal genetically mosaic mice could be produced from malignant teratocarcinoma cells after they were injected into mice blastocysts [Brinster, 1974; Mintz and Illmensee, 1975; Papaioannou et al., 1975]. Thus, the microenvironment of murine embryonic blastocysts was able to suppress the malignant phenotype of teratocarcinoma cells, which could then participate in normal organogenesis. Suppression of the neoplastic state with acquisition of specialized functions was also demonstrated for plant crown gall teratoma cells [Braun and Wood, 1976].

Experiments using murine, human, zebra-fish and chicken embryonic microenvironment models have shown the possibility of normalizing different types of cancer cells including malignant melanoma cells [Pierce et al., 1982; Podesta et al., 1984; Gerschenson et al., 1986; Hendrix et al., 2007].

Studies of nonembryonal mice tumors (several lines of rhabdomiosarcoma and Ehrlich ascitic carcinomas) transplanted into the eye anterior chamber of syngenic animals also demonstrated an increase in their differentiation and a decrease in malignancy [Shvemberger, 1986].

In 1969−1971, *in vitro*-induced differentiation of murine neuroblastoma [Schubert et al., 1969, 1971] and murine leukemic cells [Friend et al., 1971] were described. Friend murine erythroleukemia cells became a favorite model of mammalian cell differentiation in culture, with dimethylsulfoxide [Friend et al., 1971] and other substances [Leder and Leder, 1975; Preisler and Lyman, 1975; Tanaka et al., 1975; Reuben et al., 1976] as differentiation-inducing agents. The differentiation-inducing activity of dimethylsulfoxide was not limited to Friend cells and has also been shown for other mammalian cells [Kimhi et al., 1976; Kluge et al., 1976; Krystosek and Sachs, 1976], including those of humans [Collins et al., 1978]. Phorbol esters induced terminal differentiation in human promyelocytic leukemia cells [Huberman and Callaham, 1979] and in leukemic blast cells from patients with myelogenous leukemia [Koeffler et al., 1980]. Cytokines (including colony-stimulating factors) can reprogram abnormal developmental programs in leukemia, so that leukemic cells differentiate with suppression of malignancy (Ichikawa, 1969; Fibach and Sachs, 1975; Azumi and Sachs, 1977; Sachs, 1996; Lotem and Sachs, 2002a,b).

Retinoic acid was found to induce differentiation of human promyelocytic leukemia cell line HL60 [Breitman et al., 1980] and human promyelocytic leukemic cells in primary culture [Breitman et al., 1981]. Discovery of similar and improved action of all-*trans* retinoic acid (ATRA) comprised the major step forward in treatment of human acute promyelocytic leukemia [Nilsson, 1984; Daenen et al., 1986; Huang et al., 1987, 1988].

The number of differentiation inducers has been continuously increasing and their incomplete list includes all-*trans* retinoic acid (ATRA), arsenic trioxide, 5-azacytidine, bacterial lipopolysaccharide, bleomycin, bryostatin 1, butyrate, cytarabine, cytokines and colony stimulating factors, dactomucin, dimethylsulfoxide, dimethylformamide, doxorubicin, growth factor deficiency, hemin, hexamethylbisacetamide, hypomethylating agents, insulin, interferons, low glucose, mitomycin, methotrexate, phorbol esters, polybrene, prostaglandin, phosphodiesterase inhibitor, steroids, suberoylanilide hydroxamic acid, 6-thioguanine, transforming growth factor-beta, vitamin D, etc. [Reiss et al., 1986; Twardzik et al., 1989; Andreef et al., 2000; Lesczynieka et al., 2001; Lotem and Sachs, 2002a,b].

Induction of terminal differentiation *in vitro* with different agents has been accomplished with malignant cells derived from each of the three germ layers; with malignant cells originated from germ cells, embryonal tumors and tumors from definitive tissues; with cells from hematopoietic malignancies and from solid tumors; and with cells from laboratory animals and from humans. Induction of terminal differentiation of tumor cells in culture has been achieved for squamous cell carcinoma, embryonal carcinoma, colon adenocarcinoma, breast adenocarcinoma, bladder transitional cell carcinoma, prostate cancer, melanoma, neuroblastoma, erythroleukemia and promyelocytic and myelocytic leukemias [Reiss et al., 1986; Andreef et al., 2000; Lotem and Sachs, 2002a,b].

Different tumor cell lines need different inducers and/or their combinations as a cause of differentiation. For example, melanoma cells are better induced to differentiation by a combination of interferon beta and mezerein [Lesczynieka et al., 2001], and acute promyelocytic leukemia cells by a combination of all-*trans* retinoic acid and arcenic trioxide [Wang and Chen, 2008; Nowak et al., 2009]. Different compounds use alternative pathways of gene expression for inducing differentiation of tumor cells, as shown by experiments with myeloid leukemic cells [Lotem and Sachs, 2002b].

Cells from different tumors can also be induced by different agents to terminal differentiation *in vivo* in animal systems [Gootwine et al., 1982; Reiss et al., 1986; Waxman et al., 1988; Twardzik et al., 1989; Andreef et al., 2000; Lotem and Sachs, 2002a].

Tumors differ in potential for differentiation of their stem cells: embryonal carcinoma forms derivatives of all three germ layers while breast cancer stem cells form only glandular epithelium, etc. [Pierce, 1993].

Malignant cells are genetically abnormal. These abnormalities include aneuploidy, chromosome rearrangements, genetic changes in DNA, inactivation of suppressor genes, oncogenes activation etc. Genetic abnormalities of tumor cells do not preclude the induction of tumor cell differentiation, as shown by experiments with genetically abnormal tumor cells in embryonic systems [Mintz and Cronmiller, 1978], in the eye anterior chamber of syngenic animals [Alexandrova and Shvemberger, 2005], or with genetically abnormal myeloid leukemic cells [Lotem and Sachs, 2002a,b]. Reversion of the malignant phenotype of human breast cells in three-dimensional culture and *in vivo* by integrin-blocking antibodies demonstrated that the extracellular matrix and its receptors determine the phenotype of mammary epithelial cells, and that tissue phenotype is dominant over the cellular genotype [Weaver et al., 1997].

Epigenetic mechanisms control gene expression in differentiation both during normal development and in induced differentiation of tumor cells. The suppression of malignancy in genetically abnormal tumor cells occurs due to epigenetic changes. "Epigenetics wins over genetics" in induction of differentiation in tumor cells, as J. Lotem and L. Sachs formulated [Lotem and Sachs, 2002b].

The best-studied example is the mechanisms of all-*trans* retinoic acid (ATRA) induction of differentiation of acute promyelocytic leukemia (APL) cells. ATRA is a ligand to retinoic acid receptors (RARs) that bind to retinoic acid response elements (RAREs) and regulate granulocytic differentiation. The APL cells have a consistent chromosomal translocation between chromosomes 15 and 17 [Rowley et al., 1977]. The t(15;17) translocation of APL fuses the RARalpha (RARA) gene to a transcription factor PML gene [de The et al., 1990; Kakizuka et al., 1991]. The fused genes play a critical role in leukemogenesis and in determining responses to ATRA in APL, as shown in a transgenic mice model [He et al., 1998]. PML-RARA forms homodimers and recruits corepressor proteins to repress the transcription of genes essential for granulocytic differentiation by binding to RAREs in the regulatory region of these genes. PML-RARA also recruits DNA and histone-modifying enzymes such as histone deacetylases (HDAC), histone methyltransferases and DNA methyltransferases that lead to DNA hypermethylation, highly repressive chromatin conformation and transcriptional repression.

ATRA causes a conformation change and degradation of PML-RARA, and dissociation of corepressor complex. Coactivator complex composed of proteins with histone acetylase activity is recruited; this opens the chromatin structure, relieves transcriptional repression and allows differentiation to occur [Lotem and Sachs, 2002b; Collins, 2008; Wang and Chen, 2008; Nowak et al., 2009].

Arsenic trioxide (ATO) exerts selective therapeutic effects against APL by inducing degradation of PML-RARA. ATO might target the PML moiety of the fusion protein. ATO could regulate a significant proportion of genes also modulated by ATRA, but induces a deeper change of the proteome pattern, suggesting that protein modification is a major molecular mechanism of ATO. While ATRA-induced differentiation involves mostly transcriptional changes, the effects of ATO are mainly at the proteome level, which creates a molecular basis for the synergism between ATRA and ATO [Zheng et al., 2005; Wang and Chen, 2008].

Based on the discussion above and on other data, many authors came to the conclusion that cancer is a disease of differentiation [Markert, 1968; Coggin and Anderson, 1974; Pierce et al., 1978, 1982; Pierce, 1993; Sachs, 1996; Harris, 2005]. This concept will be discussed in more detail in Part 5.2. The fact that agents inducing differentiation can do so with minimal effects on normal cells and at lower concentrations than cytotoxic drugs generated an alternative approach to killing tumor cells with cytotoxic drugs, i.e. differentiation of cancer cells by changing the gene expression.

The use of agents that can induce cancer cells' differentiation for therapeutic purposes in humans has been named "differentiation therapy" [Pierce and Verney, 1961; Pierce, 1983; Sachs 1978, 1996; Reiss et al., 1986; Waxman et al., 1988; Andreef et al., 2000; Lesczynieka et al., 2001; Sell, 2004]. Interferons have been used in the therapy of melanoma to induce growth arrest and tumor differentiation. Many differentiation inducers have entered clinical evaluation in patients with leukemia or other hematological malignancies [Lesczynieka et al., 2001]. The most successful example of differentiation therapy is the treatment of acute promyelocytic leukemia with all-*trans* retinoic acid, with remission rates of up to 90% to 95%, which is considered a breakthrough in clinical oncology [Wang and Chen, 2008; Nowak et al., 2009]. Several approaches of epigenetic therapy in cancer treatment, including those using DNA demethylation drugs, are currently being explored [Berdasco and Esteller, 2010].

All these data indicate that malignancy is reversible. The genetic information responsible for the terminal differentiation of many types of tumor cells is present in these cells and may be expressed phenotypically. Tumor cells can differentiate with the loss of malignancy. The suppression of malignancy in tumor cells occurs due to epigenetic changes of gene expression. Genetic abnormalities of tumor cells do not preclude the induction of tumor cell differentiation. After tumor cells are induced to terminally differentiate, they stably lose their neoplastic growth capabilities.

Nevertheless, restoration of the normal phenotype of human breast cells in culture by integrin-blocking antibodies is reversible on removal of those antibodies [Weaver et al., 1997]. When dormant *Wnt1*-initiated cancer participates in the reconstitution of functional mammary glands, with morphologically normal ductal trees, reactivation of oncogenic Wnt signaling triggers malignant transformation of normal mammary epithelium and yields an invasive adenocarcinoma [Gestl et al., 2007]. See also the discussion of the dynamic relationship between tumors and differentiation in Part 5.2.

4.3 TUMORS PROVIDE EXCESSIVE CELL MASSES FUNCTIONALLY UNNECESSARY TO THE ORGANISM THAT COULD BE USED FOR THE ORIGIN OF NEW CELL TYPES, TISSUES AND ORGANS

Primitive multicellular organisms were small and had Type 1 embryogenesis which generated the fixed amounts of cells with predetermined fates [Davidson, 1991]. The preexisting cell types, tissues and organs of multicellular organisms have limited capacities for generation of new structures. The preexisting cell types do not seem to be capable of providing cell masses necessary for building new structures because of regulation and strict limitations imposed on the number of possible cell divisions.

It was found during serial cultivation of normal diploid cells that only limited numbers of cell divisions are possible (the so-called "Hayflick limit") [Hayflick and Moorhead, 1961]. For some organs (e.g. the pancreas) their size is limited by the number of embryonic progenitor cells [Stranger et al., 2007].

The preexisting cell types/tissues/organs already serve certain functions, which restricts their potential use for other purposes. The new structures with new functions could be generated only at the expense of the previous structures and functions.

One example of such economical use of pre-existing structures after they lose their previous functions could be beetle elytra, which have evolved all sorts of cuticular expansions, sculptures and glands [Grimaldi and Engel, 2005; Prud'homme et al., 2011].

Another example of the economical use of existing structures in evolution of multicellular organisms is the origin of auditory ossicles in mammals — the three small bones contained within the middle ear space which transmit sounds from the air to the fluid-filled labyrinth. In reptiles, the eardrum is connected to the inner ear via a single bone, the stapes, while the upper and lower jaws contain several bones not found in mammals. During the evolution of mammals one lower and one upper jaw bone (the articular and quadrate) lost their function in the jaw joint and acquired a new function in the middle ear by forming a chain of three bones (malleus, incus and stapes) that amplify sounds and allow more acute hearing, which was favored by natural selection in early mammals. The homology of the malleus and incus to the reptilian articular and quadrate is supported by embryological data and by an abundance of transitional fossils [Allin, 1975; Meng, 2003; Wikipedia, 2010; Meng et al., 2011]. The evolution of mammalian auditory ossicles is an illustration of the re-purposing of existing structures during evolution.

Duplicated morphological structures may provide material for the origin of new structures. The lower jaw of vertebrates evolved from reiterated branchial arches of the agnathans, as follows from the homology of gill arches and jaws in the primitive Devonian fish *Acanthodes* [Gregory, 1951; Raff, 1996], another example of co-option.

Ossicles remain bones; they only change their function. But from what cells and cell masses did the first bones originate? As in the case of the origin of new genes from gene duplicates that are not functionally necessary to the organism [Ohno, 1970], extra cell masses functionally unnecessary to the organism would be needed for the origin of evolutionarily new cell types, tissues and organs. What do we know about potential extra cell masses in multicellular organisms?

For mammals some continuous "post-body plan" developmental processes that include renewal of the immune system rely on dedicated set-aside stem cells [Davidson, 2006]. In nematode *Caenorhabditis elegans*, in the first stage larva, of the total 558 cells 42 postembryonic blast cells are set aside from the embryonic functions and their descendants subsequently form additional structures [Sulston et al., 1983; Davidson, 2006]. In ascidian *Ciona* the adult is built largely from set-aside larval cells. In most sea urchins and in other echinoderms the development of the adult body occurs from a specific population of embryonic set-aside cells (so-called maximum indirect development) [Davidson, 2006]. Therefore the concept of "extra cell mass" exists in embryology as "set-aside cells" that do not participate in the embryonic specification and differentiation functions, but have a function in the future development [Davidson et al., 1995; Davidson, 2006].

Asexual reproduction and regeneration are connected with generation of considerable cell masses that participate in formation of the missed "body-plan" structures. These processes may involve undifferentiated stem cells (neoblasts) (e.g. in *Planaria* [Sanchez Alvarado and Newmark, 1998]), or dedifferentiated cells of the covering epithelium (e.g. in *Annelida* [Kharin et al., 2006]). Planarian neoblasts are pluripotent adult stem cells that underlie regeneration, produce cells that differentiate into neuronal, intestinal, and other known cell types and are distributed throughout the body [Wagner et al., 2011]. Wnt signaling is involved and promotes tail regeneration. Polarized *notum* activation at wounds inhibits Wnt function and promotes planarian head regeneration [Petersen and Reddien, 2011].

The question remains whether these or similar processes, which are under control within the "body-plan" feedbacks and regulatory networks, can significantly contribute to the origin of evolutionarily new structures and organs.

On the other hand, tumors could provide such excessive cell masses to evolving multicellular organisms for morphological novelties. The formation of excessive cell masses functionally unnecessary to the organism is considered a major feature of tumors, especially in invertebrates and lower vertebrates [Sparks, 1985; Sinderman, 1990; Grizzle and Goodvin, 1998]. To quote R.J. Roberts, "There are many definitions of neoplasia. Literally it means new growth, but probably the most specific definition, embracing all of its facets is as follows: the multiplication of cells in an aberrant fashion, which results in *excessive numbers of cells*, whose growth is often uncoordinated, and persists after the stimulus which initiated it has ceased to exert its effect" [Roberts, 2001][3].

To be evolutionarily meaningful, the cellular proliferative process should be reproduced in subsequent generations, and it is known that tumors could be genetically or epigenetically determined (see Part 6.1 for more discussion of hereditary tumors).

4.4 TUMORS AS ATYPICAL ORGANS/TISSUES THAT MAY EVENTUALLY EVOLVE INTO NORMAL STRUCTURES

Solid tumors grow as solid cell masses. In this respect, they differ from leukemias and ascites tumors. The structure of solid tumors resembles the structure of normal tissues. A solid tumor consists of parenchyma, i.e. neoplastic cells themselves, and the stroma in which neoplastic cells are dispersed. Stroma includes connective tissue, blood vessels and inflammatory cells. Increased vascular permeability is important for tumor stroma pathogenesis. It is connected to vascular permeability factor/vascular endothelial growth factor (VPF/VEGF), which is a multifunctional cytokine critical to tumor angiogenesis and stroma formation [Connolly et al., 2000].

There are similarities between tumor stroma generation and wound healing: VPF/VEGF expression is upregulated, a local increase in vascular permeability takes place and infiltration of new blood vessels and connective tissue cells occurs, leading to the development of connective tissue. The difference is that VPF/VEGF expression and vascular hyperpermeability in tumors persist indefinitely, which leads to the development of poorly supportive connective tissue of malignant tumors and necrosis. The presence of necrosis usually distinguishes malignant tumors from benign tumors [Dvorak, 1986; Connolly et al., 2000; Schafer and Werner, 2008].

Benign tumors consist of well-differentiated cells that resemble the corresponding cells of normal tissue. Benign tumors expand in all directions and have an enveloping connective tissue rim that may serve as a capsule that separates the tumor from surrounding normal tissues.

---

3. Italics are mine — A.K.

In malignant tumors, neoplastic cells are less differentiated (anaplastic) and have altered cytological features (altered polarity, increased nuclear to cytoplasmic ratio, pleomorphism, etc.). Neoplastic cells and stroma have disorganized orientation. Malignant tumors lack a capsule; their growth is invasive to surrounding tissues. Malignant carcinomas that have not extended through the underlying basement membrane are referred to as *in situ* carcinomas [Connolly et al., 2000].

Tumors have a hierarchical nature. They consist of a mixture of cell types at different stages of differentiation, similar to those in normal organs. At least two different populations of cells can be recognized: a minority of undifferentiated cancer stem cells (CSCs) and a majority of "derived cells," more differentiated and with a limited life span. The derived cells would include transit-amplifying cells and terminally differentiated cells. More cellular heterogeneity is added by different locations within a tumor mass and by the presence of non-tumor cells, i.e. inflammatory cells, tumor-associated fibroblasts, endothelial cells, pericytes, stem and progenitor cells of the tumor stroma and immature myeloid cells [Pierce et al., 1977; Hanahan and Weinberg, 2000, 2011; Alison et al., 2010; Maenhaut et al., 2010]. Normal ancillary cells of different types play important roles in supporting tumor cells proliferation [Hanahan and Weinberg, 2000, 2011].

Like normal stem cells, CSCs seem to depend on vascular niches and factors secreted there. On the other hand, tumor angiogenesis is stimulated by CSC-secreted VEGF. Thus, there is a positive feedback between CSCs and angiogenesis/endothelial cells, which is characteristic of a symbiotic relationship [Eyler and Rich, 2008; Maenhaut et al., 2010].

All this evidence suggests that tumors may be considered as complex heterogeneous tissues or organ-like systems [Hanahan and Weinberg, 2000, 2011; Reya et al., 2001; Brabletz et al., 2005; Boman and Wicha, 2008; Eyler and Rich, 2008], or "quasi tissues," that contain both stem cells and differentiated cells, or complex tridimensional tissues with phenotypically and functionally heterogeneous cells [Dalerba et al., 2007; Morrison and Spradling, 2008].

It was suggested that carcinomas can be viewed as atypical organs [Vermeulen et al., 2008]. This view is based on the studies of colon CSCs. The data support the concept that single CSC can self-renew and reconstitute a complete and differentiated carcinoma. It was proven that the CD133+ /CD24+ cell population contains clonogenic potential and that multilineage differentiation is intrinsic to CSCs, i.e. the differentiation heterogeneity in colorectal carcinomas is a clonal trait. According to the model, carcinomas are atypical organs that have a functional stem cell compartment in which crucial mechanisms for homeostatic control are lost but other important characteristics are present [Vermeulen et al., 2008].

Even though the tumor cells are organized in organ-like structures, they still are not functionally necessary to the organism. Atypical tumor organs may be used by natural selection for the origin of normal organs if they acquire regulated function and if the tumor-bearing organisms survive long enough to leave progeny. The placenta could be an example of such a regulated tumor-like organ (see Parts 5.6 and 7.5 for more discussion of the placenta as a regulated tumor).

4.4.1 Morphogenetic Potential of Tumors May Be Used in the Origin of Morphological Novelties and Diversity

Tumors may participate in morphogenetic processes, as in the case of horny unpigmented papillomas or "cutaneous horns" [Willis, 1967]. Peculiar caps formed on the heads of some fishes proved, at close examination, to be benign papillomas [Zabezhinsky et al., 2010; Kozlov et al., 2012] (see Part 11.2 for more discussion of goldfish hoods).

It is possible that the unique outgrowths formed in different species have a papillomatous or tumor-like nature, for example, the muzzle of the warthog *(Phacochoerus aethiopicus)*, regarded as its systematic trait; the cock's comb and gill; the turkey-cock's jowl; the thoracic expansions that evolved in horn beetles; etc.

The resemblance of tumor growth to the growth of horns in deer, cattle and rhinoceros was noticed early by students of comparative oncology [Bland-Sutton, 1890, as cited in Huxley, 1958].

Beetle horns are tubular cuticular projections from the body wall. During development horns form from compact discs of epidermal cells that proliferate during the late larval period and evert during the pupal molt [Moczek et al., 2006; Shubin et al., 2009]. Although horn development resembles the development of insect appendage imaginal discs, they do not arise from imaginal discs but from the novel discs — the new regions of epidermis which at a certain period in evolution began to behave like imaginal discs [Emlen et al., 2007]. For my consideration it is interesting that prepupal horn primordia in different *Onthophagus* species (see Figure 4 in [Moczek et al., 2006]) look like papillomas [M.A. Zabezhinskiy, personal communication]. Four appendage patterning genes (*exd*, *hth*, *dac* and *Dll*) are expressed in prepupal horn primordia [Moczek et al., 2006], suggesting that beetle horns are the product of the co-option and deployment of an appendage patterning program at novel anatomical sites [Shubin et al., 2009] and in papilloma-like outgrowths.

An interesting example is the treehopper (*Membracidae*) "helmet," characterized by high morphological diversity. It was recently discovered that the "helmet" is an extra wing-like appendage, a wing serial homolog on the first thoracic segment, an unprecedented situation in 250 Myr of insect evolution [Prud'homme et al., 2011]. The authors have shown that the helmet arose by escaping the ancestral repression of wing formation on the first thoracic segment by a member of the *Hox* gene family. They interpret this phenomenon as an example of expression of ancestral developmental potential, the part of which is expression of transcription factor Nubbin. The exceptional morphological diversification of the helmet is hypothesized to be due to escape from the stringent functional requirements [Prud'homme et al., 2011].

It should be pointed out, though, that the escape from ancestral repression and the expression of ancestral developmental potential do not lead to the development of the third pair of wings on the first thoracic segment, as described for early insect fossils [Grimaldi and Engel, 2005], but to highly diverse structures that mimic thorns, animal droppings and ants and that sometimes resemble tumors.

In this respect, it is interesting to think of the "embryonic remnants" theory of tumors, which in evolutionary perspective may have a relationship to the above example. The cellular outgrowth without functional requirements could be considered a tumor. The "embryonic remnants" theory considers tumors as an expression of unrealized developmental potential. On the other side, the *Hox* genes may demonstrate oncogenic potential. The mixture of embryonic and tumor processes may be used in evolution and lead to changes in developmental programs and to the origin of new features and morphological novelties. It may be true for more than the above examples and is discussed in more detail in the next chapter.

Tumors and/or tumor-like processes could play a general and important role in the evolution of ontogenesis. The ability of tumor cells to differentiate with concomitant loss of malignancy, in combination with their ability to express genes that are not expressed in normal tissues and with other tumor features discussed above, may result in the emergence of new cell types, tissues and organs in evolution.

Tumors Might Participate in the Evolution of Ontogenesis

5.1 TUMORS AND NORMAL EMBRYOGENESIS

The idea that cancer is the problem of developmental biology and that tumors somehow relate to embryonic development originated in the nineteenth century. Rudolf Virchow was one of the first to discuss this relationship:

"... Genuine new-formation goes on, and that in accordance with the same law, which regulates embryonic development.

This law of correspondence between embryonic and pathological development was ... laid down by Johannes Muller, who continued the investigations commenced by Schwann" [Virchow, 1860].

According to C. Oberling, as early as 1829 French investigators Lobstein and Recamier attributed the origin of tumors to the proliferation of embryonal cells that had persisted into adulthood. This idea was supported and further developed by different authors throughout the nineteenth century. Johannes Muller, Paget, Remak, Durante and Cohnheim were mentioned in this connection [Oberling, 1946; cit. in Gurchot, 1975].

F. Durante and J. Cohnheim introduced the "embryonal rest" or "embryonic remnants" theory of cancer [Durante, 1874; Cohnheim, 1877, 1889]. According to this theory, more embryonic cells are produced than necessary for formation of adult tissues and organs. The excess embryonic cells may stay dormant but capable of neoplastic development under appropriate conditions. In a biologically different line of reasoning, John Beard developed the "trophoblast theory" of cancer, which suggested that cancer originates from primordial germ cells that fail to complete their embryonic migration to gonads [Beard, 1902].

Some modern authors think that the "embryonic rest" concept of Durante and Cohnheim is basically correct and that there is considerable similarity of this concept with the modern theory of cancer stem cells [Pierce, 1993; Sell, 2004; Hendrix et al., 2007]. In particular, it is supported by the expression of the variety of carcinoembryonic antigens in tumors discussed above.

Evolution by Tumor Neofunctionalization.

Brain tumors and embryonic tumors developing after infancy are explained by the persistence of cell rests or developmental vestiges [Glazunov, 1947; Bolande, 1967; Sell, 2004; Meizner, 2011].

The cancer "trophoblast theory" of Beard is supported by frequent production of chorionic gonadotropin and other trophoblastic hormones by cancer cells of different types and by the expression of cancer/testis antigens and germ cell genes in tumors [Gurchot, 1975; Acevedo et al., 1995; Simpson et al., 2005; Janic et al., 2010]. The theory of Beard was "revisited" and restated in a modified form in modern terms by Charles Gurchot in 1975 [Gurchot, 1975].

The modern theory interrelating cancer, cell differentiation and embryonic development is the "stem cell" theory of cancer. Many aspects of this theory originated from studies of teratocarcinomas [Pierce et al., 1978; Pierce, 1993] and leukemias [Sachs, 1996]. The stem cell theory of cancer eventually led to the successful therapy of acute leukemia, as discussed in Part 4.2. The last two decades have produced a wealth of data on stem cells (SCs). The current state of knowledge includes the notions of embryonic stem cells (ESCs), adult or somatic stem cells (SSCs), and cancer stem cells (CSCs) [Alison et al., 2010; Maenhaut et al., 2010].

Tumors contain a subpopulation of malignant cells with stem cell properties – CSCs. The cancer stem cell hypothesis suggests that tumorigenesis begins in normal adult stem cells or progenitor cells that have recently descended from them [Clarke et al., 2006]. Mutations in the DNA of normal stem cells, or epigenetic changes in normal stem cells [Feinberg et al., 2006], appear to be the initiating event in many cancers. CSCs may also arise from more restricted progenitor cells. The CSCs are similar to normal stem cells in their capability for self-renewal. They contain multilineage differentiation capacity, although aberrant, and give rise to a hierarchy of progenitor and differentiated cells. CSCs also differ in many respects from the ESCs and SSCs. Most importantly, CSCs lack developmental control of ESCs and homeostatic control of SSCs populations. That is, tumors lack the functional feedback loop that regulates normal stem cells.

CSCs cause tumors. CSCs are often referred to as tumor-propagating cells (TPCs), reflecting their capability to cause the development of tumors when xenografted into immunodeficient mice. TPCs are the most tumorigenic cells of tumor cell population. TPCs have defining markers discussed above in Part 4.1 and can be assessed in *in vitro* and *in vivo* assays [Vermeulen et al., 2008; Alison et al., 2010; Hubbard and Gargett, 2010; Maenhaut et al., 2010]. The function of CSC or TPC markers is not fully understood, but it is now recognized that some genes may cause tumors by activating early developmental pathways associated with programming for multipotency, a feature of stem cells [Janic et al., 2010; Stange et al., 2010; Wu and Ruvkun, 2010]. First described morphologically [Pierce et al., 1977], the CSCs were physically isolated using the cell surface markers from leukemia [Lapidot et al., 1994; Bonnet and Dick, 1997] and afterwards from solid tumors [Al-Hajj et al., 2003; Singh et al., 2003; Visvader and Lindeman, 2008]. After implantation of CSC-enriched populations *in vivo* tumors were generated, with tumor cell populations no longer enriched for CSCs. That is, CSCs self-renew and also produce non-CSC, more differentiated progeny (reviewed in [Gupta et al., 2009]).

The signaling pathways that participate in regulation of self-renewal of normal SCs in different organs include Wnt, Sonic hedgehog (Shh) and Notch. The Wnt pathway regulates self-renewal in hematopoietic, epidermal, neural and gut stem/progenitor cells; Shh − in hematopoietic, epidermal, neural, and germ line stem/progenitor cells; and Notch − in hematopoietic, epidermal, neural, gut and germ line stem/progenitor cells. These pathways participate in the normal developmental process. For example, Notch signaling acts in a common pathway with asymmetric cell division, promoting the differentiation switch during mouse skin development [Williams et al., 2011]. Wnt, Shh and Notch are among the fundamental signaling pathways that drive pattern formation in animals [Gerhart, 1999; Barolo and Posakony, 2002]. But Wnt, Shh and Notch also participate in tumor progression when deregulated [Taipale and Beachy, 2001; Beachy et al., 2004]. Their deregulation causes deregulation of SCs' self-renewal which leads to neoplastic growth. Mutations in the Wnt pathway are connected with colon carcinoma, epidermal tumors, medulloblastoma and chronic myeloid leukemia; in Shh with medulloblastoma, breast cancer and basal cell carcinoma; and in Notch with T cell leukemias, chronic lymphocytic leukemia, medulloblastoma, mammary tumors, colon cancer, melanoma and head and neck squamous cell carcinoma (HNSCC) [Reya et al., 2001; Balint et al., 2005; Reya and Clevers, 2005; Hatsell and Frost, 2007; Zhao et al., 2007; Boman and Wicha, 2008; Charafe-Jauffret et al., 2008; Fan and Eberhart, 2008; Visvader and Lindeman, 2008; Sikandar et al., 2010; Agrawal et al., 2011; Puente et al., 2011; Stransky et al., 2011]. *NOTCH1* appears to be an important tumor suppressor gene in HNSCC [Agrawal et al., 2011; Stransky et al., 2011].

Other genes originally discovered as determinants of development demonstrate oncogenic potential.

The RUNX family of transcription factors plays an important role in normal development in *D. melanogaster, C. elegans* and in mammals. RUNX1 (or AML1) participates in hematopoietic development, RUNX2 is connected with osteoblast differentiation and bone ossification, and RUNX3 is involved in development of the gastrointestinal tract. All of them also contribute to tumorigenesis as oncogenes (*RUNX1* and *RUNX2*) or a tumor suppressor gene (*RUNX3*) [Lund and van Lohuizen, 2002].

In vertebrates, an embryonic morphogen Nodal belonging to the TGFβ superfamily induces embryonic axis formation, patterns the nervous system, induces the mesoderm and endoderm and determines left−right asymmetry.

Studies of reprogramming metastatic tumor cells with embryonic microenvironments have led to the discovery of the reciprocal influence of melanoma cells on embryonic development of zebrafish: when the melanoma cells were injected towards the animal pole the embryos developed an abnormal anterior outgrowth, and when the cells were injected near the yolk margin a duplication of the body axis occurred. These observations led to the discovery that metastatic melanoma cells secrete Nodal. Thus Nodal maintains the pluripotency of human embryonic stem cells and sustains melanoma tumorigenicity and plasticity [Schier, 2003; Topczewska et al., 2006; Hendrix et al., 2007].

Retinoic acid, its receptors (RARs) and retinoid binding proteins play an important role in vertebrate morphogenesis and differentiation [Ruberte et al., 1990, 1991; Ruiz I Altaba and Jessel, 1991; Cvekl and Piatigorsky, 1996]. The t(15;17) translocation of acute promyelocytic leukemia (APL) fuses the RARalpha (RARA) gene to a transcription factor PML gene [de The et al., 1990; Kakizuka et al., 1991]. The fused genes play a critical role in leukemogenesis, as discussed above in Part 4.2.

Heat shock protein Hsp90 is a molecular chaperone assisting the correct folding of cellular proteins. By supporting a variety of signal transducers, Hsp90 is involved in several developmental pathways. As experiments have shown, the role of Hsp90 may consist in buffering the widespread variation affecting morphogenetic pathways. When Hsp90 buffering is compromised, cryptic variants may be expressed. Hsp90 thus may serve as a capacitor for morphological evolution [Rutherford and Lindquist, 1998]. At the same time, studies of oncogenic tyrosine kinase inhibitors have shown that Hsp90-src heteroprotein is required for *src*-mediated transformation and may be a target of anticancer drugs [Whitesell et al., 1994]. Since then the analysis of Hsp90-containing complexes demonstrated that the client proteins in these complexes include members of many signaling pathways which may be deregulated in cancer, and many promising anticancer drugs − Hsp90 inhibitors − are in clinical trials [Darby and Workman, 2011].

Homeobox genes, the major development regulators, are involved in tumors. Oncogenic action of the *Hox-2.4* gene is connected to its ability to impede the IL-3-driven terminal differentiation of myeloid cells [Perkins et al., 1990]. The homeobox *Hox-11* gene is involved in translocations leading to acute T cell leukemia [Rabbitts, 1991]. HOXB7 constitutively activates basic fibroblast growth factor in melanomas [Care et al., 1996]. Deregulated expression of *Hox* genes has been described in many carcinoma cell lines and solid tumors [Peverali et al., 1990; Cillo et al., 1992, 2001; Abate-Shen, 2002]. The role of homeobox genes in tumors (*HOX, EMX, PAX, MSX* gene families and other divergent homeobox genes) is reviewed in [Cillo et al., 1999, 2001]. It was generalized that homeobox genes that are normally expressed in undifferentiated cells are upregulated in cancer, and those that are normally expressed in differentiated tissues are downregulated in cancer [Abate-Shen, 2002].

Some authors use different logical order — they say that many pathways classically associated with cancer may also regulate normal stem cell development [Reya et al., 2001]. It is claimed that networks of protooncogenes and tumor suppressors have evolved to coordinately regulate stem cells [Kozlov, 1996; Pardal et al., 2005].

There are indeed plenty of data on the connection of protooncogenes and tumor-suppressor genes with normal development. Originally, studies with *Drosophila* and *Caenorhabditis elegans* have shown that protooncogene homologs (in particular, homologs to receptor protein tyrosine kinases) participate in signal transduction pathways connected with development. Ras protein has been shown to act in the same pathway downstream from the tyrosine kinase [Hoffman et al., 1992]. Germline mutations in the Ras-Raf-MEK-ERK pathway cause several developmental disorders, including Noonan, Costello and cardio-facio-cutaneous syndromes [Schubbert et al., 2007]. It turned out that protooncogenes and their homologs affect a variety of developmental processes in many organisms, including mice, flies, worms and yeast. The striking examples of this kind are provided by *Wnt* genes [Nusse and Varmus, 1992]. *Wnt-1* was discovered as a protooncogene responsible for mammary tumors induced by mouse mammary tumor virus [Nusse and Varmus, 1982]. Activation of the Wnt signal transduction pathway plays a role in a variety of human and mice tumors [Nusse and Varmus, 1982, 1992; Peifer and Polakis, 2000; Polakis, 2000; Reya and Clevers, 2005; Fan and Eberhart, 2008]. On the other hand, the *Wnt* gene family encodes a group of cell-signaling molecules that participate in vertebrate and invertebrate development. The Wnt protein sequences have been conserved during a billion years of evolution, suggesting common mechanisms of action. Mutations in *Wnt* genes or Wnt pathway components lead to specific developmental defects [Nusse and Varmus, 1992; Parr et al., 1993; Cadigan and Nusse, 1997; Nusse, 2001; Kusserow et al., 2005].

The *HER2* oncogene regulates the normal mammary stem/progenitor cell population, and its overexpression drives mammary carcinogenesis, tumor growth and invasion [Korkaya et al., 2008]. Normal roles in differentiation are also established for the oncogenes encoding basic-helix-loop-helix and LIM proteins and for the retinoic acid receptor involved in translocations of acute T cell leukemia [Rabbitts, 1991]. The expression of src affects the synthesis of differentiated cell products in a variety of cell types. For example, in rat PC12 phaeochromocytoma cells, which are used as a model to study neuronal differentiation and the mechanism of action of nerve growth factor (NGF), expression of v-src has an inductive effect on differentiation that resembles the action of a NGF [Alema et al., 1985].

The phenotypic analysis of mice carrying germline mutations in protooncogenes provided genetic evidence for the important role that these genes play in mammalian development and differentiation [Forrester et al., 1992].

Germline mutations of *PTEN* and *LKB1* tumor suppressor genes cause related Cowden disease (CD) and Peutz-Jeghers syndrome (PJS), respectively [Smith and Ashworth, 1998]. CD and PJS are autosomally dominantly inherited diseases that share clinical features, such as multiple tumors. *PTEN* and *LKB1* tumor suppressor genes display largely overlapping expression patterns during embryonic development [Lukko et al., 1999]. According to other authors, *PTEN* plays a crucial role in embryogenesis, in the pathogenesis of CD, Lhermitte-Duclos disease (LDD) and Bannayan-Zonana syndrome (BZS), characterized by tumor susceptibility and developmental defects. *PTEN* plays a role in tumor suppression through its ability to control cellular differentiation and anchorage-independent growth. The authors consider *PTEN* as a tumor suppressor essential for embryonic development [Di Cristofano et al., 1998].

The tumor suppressor gene *Brca1* is required for embryonic cellular proliferation in the mouse [Hakem et al., 1996] and rat [Korhonen et al., 2003]. It controls proliferation and differentiation of neural progenitor cells [Gowen et al., 1996; Korhonen et al., 2003]. There are more examples of the participation of protooncogenes and tumor-suppressor genes in the development of brain, discussed below in Parts 5.5.1 and 5.5.2.

Stem cells, protooncogenes, homeobox-encoding genes, zebrafish Nodal and other findings discussed above indicate the convergence of tumorigenic and embryonic signaling pathways. Canonical signaling pathways are fundamentally involved in development of both normal organs and tumors of various histogenesis, i.e. atypical organs. Normal stem cells and tumor stem cells are regulated by common networks that include protooncogenes and tumor suppressors. This evidence suggests the fundamental connection of tumors and embryonic development and the possibility that tumors might participate in evolution of ontogenesis.

5.2 TUMORS AS DISEASE OF DIFFERENTIATION

Although CSCs have differentiation capabilities, they fail to differentiate correctly. This failure may lead to tumor formation. Tumor cells may be induced to differentiate experimentally, and patients with neoplasms could be treated with differentiation therapy, as discussed above in Part 4.2. This means that the differentiation process is seriously impaired in tumors, and cancer in many respects may be considered as disease of differentiation. Many authors contributed to this concept [Markert, 1968; Coggin and Anderson, 1974; Pierce et al., 1978, 1982; Pierce, 1993; Sachs, 1996; Harris, 2005].

Clement L. Markert in his classical paper "Neoplasia: A Disease of Cell Differentiation" [Markert, 1968] was probably one of the first to introduce the terminology. According to Markert, aberrations in development are expressed in neoplasms; abnormal combination or excess of normal cell properties, i.e. cell division, cell migration and the metabolic pattern or bio-chemical activity of the cell, is a feature of neoplasms; all variations in any of these three fundamental characteristics can be attributed to the aberrant programming of gene function in neoplasms. Cancer is a special expression of abnormal programming of gene function during cell differentiatin [Markert, 1968].

In 1974, J.H. Coggin, Jr. and N.G. Anderson published a comprehensive paper on cancer, differentiation and embryonic antigens [Coggin and Anderson, 1974]. In this paper, the authors presented the detailed review of the concept that cancer is a disease of the mechanism(s) of differentiation. Their standpoint is that all major changes associated with tumors are due to changes in programming of normal genes. The authors came to the conclusion that all essential characteristics of neoplasia are found in embryonic cells and that selective re-expression of some of the genes involved in early development occurs in cancer [Coggin and Anderson, 1974].

Gordon Barry Pierce was among pioneers of research on the boundary between cancer, cell differentiation and embryonic development. He was the first to discover that tumor cells of teratocarcinomas could differentiate with the loss of malignancy [Pierce and Dixon, 1959]. At that time, these were revolutionary data that established a new paradigm in oncology. Due to the research of Pierce and his colleagues it became clear that undifferentiated appearance of malignant cells was not due to "dedifferentiation" but due to overproduction of undifferentiated cells that have a limited potential for differentiation [Pierce et al., 1977]. According to the concept in which Pierce summarized the vast data obtained on different models (teratocarcinoma, testicular tumors, breast cancer, the squamose cell carcinomas of the skin and adenocarcinomas of the colon), in tumors there is a gross overproduction of undifferentiated stem cells as compared to stem cells that differentiate. During carcinogenesis normal undifferentiated stem cells give rise to undifferentiated malignant stem cells. Normal regulated stem cells proliferate and produce the correct number of differentiated cells, whereas malignant stem cell proliferation leads to formation of excessive mass of undifferentiated progeny cells. The undifferentiated mass is the result. Tumor cells could be regulated in the appropriate "embryonic field" − a temporal and anatomical site in the embryo that abrogates the malignancy of tumors derived from that field's normal lineages. The ultimate cure for cancer will be through the re-regulation of malignant stem cells to benign stem cells using the principles of embryonic induction and the growth factors' action [Pierce et al., 1978; Arechaga, 1993; Pierce, 1993].

A similar concept was developed as a result of studies of hematopoiesis and leukemia. Hematopoiesis generates blood cells of different lineages during ontogenesis. Abnormalities in the normal developmental program lead to blood cell diseases including leukemia. Malignancy can be suppressed in certain types of leukemic cells by inducing differentiation with cytokines that regulate normal hematopoiesis or with other compounds that use alternative differentiation pathways [Sachs, 1996].

On the basis of studies of the oncofetal markers in hepatocellular carcinomas (HCCs) described by Harold Morris [Morris and Wagner, 1968; Morris and Meranze, 1974], Van Rensselaer Potter also came to the conclusion that oncology is blocked ontogeny. He suggested that the expression of fetal liver cell markers in HCCs was due to a block in the maturation of immature liver cells that causes cancer [Potter 1978, 1981; Sell and Leffert, 2008]. The modern state of knowledge about hepatocarcinogenesis is that the maturation arrest of cells at various stages of differentiation in a hierarchical cell lineage may best explain the various types of liver cancer [Sell and Leffert, 2008].

Retinoblastoma tumors arise from clones that escape terminal differentiation-associated growth-arrest. Thus the route to retinoblastoma transformation is in bypassing terminal differentiation [Chen et al., 2004].

Henry Harris discussed the concept that cancer is a disease of differentiation in a review which he called "A long view of fashions in cancer research" [Harris, 2005]. Harris came to conclusion that cancer is not caused by direct action of oncogenes, not fully explained by the impairment of tumor suppressor genes, not set in motion by mutations controlling the cell cycle, not governed by the dependence of malignant tumors on an adequate blood supply and not triggered by a failure of programmed cell death. But he thinks there is strong evidence that cancer is the disease of differentiation. In particular, he finds support for this argument in the works of B.M. Mechler and his colleagues on *Drosophila melanogaster*. In *Drosophila*, a series of genes (about 40) have been identified which, when mutated, produce tissue-specific tumors in a wide range of tissues. Homozygous recessive mutations in these genes interrupt the differentiation of adult primordial cell types and lead to an uncontrolled, invasive cell growth resulting in the premature death of the mutant animal. The *lethal(2)giant larvae (l(2)gl)* gene was the best studied in this respect. Homozygous *l(2)gl* mutations cause the development of malignant tumors in the brain and in the imaginal discs. Mutations in other loci cause benign neoplasms [Mechler et al., 1985, 1989; Mechler, 1994; Harris, 2005].

Harris stresses that tumors occurred at specific times and specific sites when mutations blocked the execution of critical steps in the process of normal differentiation. He emphasizes that *Drosophila* experiments are dealing with a manageable number of specific genes, active in different tissues, and that nothing of comparable precision has yet been achieved in mammalian cells [Harris, 2005].

The mutations that produce tumors are loss of function mutations, and the function lost is an essential part of the normal differentiation program of the affected tissue. Introduction back into *l(2)gl* mutant animals of the cloned DNA from normal *l(2)gl* locus by P-element-mediated transformation restored normal development [Mechler et al., 1989].

Similar data about the dynamic character of the relationship between tumors and differentiation were obtained on mammalian cell hybrids. In hybrids formed by fusing a range of different malignant tumor cells with normal diploid fibroblasts or diploid keratinocytes, malignancy, as measured by progressive growth *in vivo*, is suppressed when the composite cell retains the ability to execute the differentiation program of the fibroblast or the keratinocyte (i.e. production of a collagenous extracellular matrix or involucrin protein, respectively). When due to chromosome segregation in hybrid cells the capability to synthesize the differentiation proteins is lost, the malignant phenotype regains, and the segregant cells again acquire the capability to grow progressively *in vivo* [Harris, 1985, 2005; Harris and Bramwell, 1987].

The data obtained on *Drosophila* and mammalian cell hybrids add complementary evidence to the data on the ability of tumor cells to differentiate with the loss of malignancy, discussed above in Part 4.2. In cell hybrids, the loss of malignancy is dependent on continuous presence of differentiation product. In *Drosophila*, cellular overgrowths are caused by mutations leading to the absence of differentiation products.

Thus the relationship between tumors and differentiation is dynamic. Tissue-specific functions are connected with their corresponding genes/differentiation products, and the loss of functions (e.g. due to mutations) is connected with tumors. The lack or disruption of a functional regulatory feedback loop may be the cause of tumors. On the other hand, the role of functional regulatory feedbacks appears as determinative in the differentiation of tumors when it occurs, and could be used in the evolutionary origin of new cell types/tissues/organs from tumors (see discussion of placenta as regulated tumor in Part 7.5).

5.3 THE EPITHELIAL TO MESENCHYMAL TRANSITION (EMT) OCCURS IN NORMAL AND NEOPLASTIC DEVELOPMENT

The epithelial to mesenchymal transition (EMT) is the formation of mesenchymal cells from epithelia in different parts of an embryo or *in vitro* in epithelial tissue explants. EMT is the process of transdifferentiation, during which the epithelial phenotype transforms into mesenchymal phenotype. During EMT the cell–cell adherence reduces due to repression of E-cadherin and other cell–cell junction molecules. Expression of genes coding for mesenchymal filament protein vimentin and other mesenchymal markers is activated together with multiple EMT-inducing transcription factors such as Snail or Twist. This is associated with the change of cellular shape to fibroblast-like appearance and with increased cell motility. EMT is not irreversible. The reverse process – mesenchymal to epithelial transition (MET), through which mesenchymal cells re-establish epithelial phenotype – also occurs. Several rounds of EMT and MET take place during normal development. As a result of initial EMT, the primary mesenchyme is formed. MET of primary mesenchyme generates multiple mesoderm structures.

These structures undergo a second wave of EMT (secondary EMT) generating mesenchymal cells that differentiate into specific cell types. An example of tertiary EMT would be formation of the cushion mesenchyme in the heart. Thus EMT and MET play important roles in differentiation and formation of many tissues and organs [Thiery, 2003; Hugo et al., 2007; Thiery et al., 2009].

A wealth of data accumulated proves that EMT and MET also participate in neoplastic development, i.e. in carcinoma progression and metastasis (reviewed in [Thiery and Sleeman, 2006; Hugo et al., 2007; Chaffer and Weinberg, 2011]). Decreased E-cadherin expression and expression of many markers of mesenchymal phenotype have been described in a variety of cell lines from breast and prostate cancer. Vimentin expression in cervical carcinomas was shown to be associated with invasive and migratory potential [Gilles et al., 1996].

The same signaling pathways that regulate developmental EMT are also activated during tumor progression [Thiery and Sleeman, 2006]. Accumulation of nuclear β-catenin (the evidence of Wnt signaling pathway activation) was demonstrated at the invasive front and in tumor cells migrating into stroma in colorectal cancer. The highest accumulation of nuclear β-catenin was found in dissociated, dedifferentiated tumor cells that have undergone EMT. On the other side, most metastases re-express E-cadherin and have lower expression of β-catenin, indicating an MET reversal in established metastases [Brabletz et al., 1998, 2005; Hlubek et al., 2007].

The migrating cancer stem cell concept, that combines stem cell and EMT concepts, was suggested to explain the malignant tumor progression, i.e. the existence of mobile cancer stem cells which transiently develop from stationary CSCs through EMT [Brablez et al., 2005].

Normal and cancer stem cells express EMT markers. The epithelial-mesenchymal transition generates cells with properties of cancer stem cells [Mani et al., 2008; Morel et al., 2008; Campbell Marota and Polyak, 2009; Evdokimova et al., 2009]. Epithelial carcinoma cells often express both epithelial and mesenchymal markers, i.e. represent a phenotypic state that is not encountered in normal tissues [Santisteban et al., 2009; Chaffer and Weinberg, 2011]. As emphasized by the authors, tumor-associated EMT may be incomplete or aberrant [Santisteban et al., 2009].

The cancer stem cells (CSCs) differentiate into non-CSCs, as discussed in Parts 4.2, 5.1 and 5.2. The concept of interconvertibility of CSC and non-CSC populations via a process which is based on greater plasticity of tumor cell populations and closely allied to EMT was suggested [Gupta et al., 2009]. This concept may explain the varying proportions of CSCs reported in different papers. The possibility of CSC differentiation into non-CSCs and vice-versa is important for my consideration of evolution by tumor neofunctionalization.

Taken together, the different sets of data and concepts about stem cells, cancer as disease of differentiation, signaling pathways and EMT provide understanding of fundamental and dynamic links between neoplastic and normal embryonic development. This fundamental dynamic linkage existed throughout evolution and might have played a role in the evolution of ontogenesis. Throughout the evolution of multicellular organisms, their normal development was accompanied by different kinds of neoplasms, and evolution might use this connection for further elaboration of ontogenesis, or evo-devo.

5.4 TUMORS, EVO-DEVO AND ADDITION OF FINAL STAGES IN THE EVOLUTION OF ONTOGENESIS

The interrelation between phylogeny and ontogeny has interested scientists since Aristotle (reviewed in [Gould, 1977]). Evolutionary biologists realized that it was the whole ontogenesis that evolved and was selected for fitness, not just the adult organisms. The outstanding generalizations on this relationship have been formulated, including the laws of development by Karl Ernst von Baer [von Baer, 1828] and Haeckel's biogenetic law [Haeckel, 1866]. Russian evolutionist A.N. Severtsov and his school defined the following major mechanisms in the evolution of ontogenesis, or modes of phylembryogenesis, as he called them: archallaxis, when embryogenesis was changed from the earliest stages (e.g. the origin of the third germ layer in triploblastic animals); deviation, when the changes were introduced in the intermediary stages of embryogenesis (e.g. the intercalation of the planktonic larval forms within the ontogenesis of their lineages [Wolpert, 1999; Valentine and Collins, 2000]); and anaboly, when characters were added at terminal stages of ontogenesis, i.e. addition of the final stages of morphogenesis [Severtsov, 1927, 1935, 1949; see also Gould, 1977 for review].

Parallelisms in the evolution of tissues in different groups of organisms [Zavarzin, 1934, 1953] and the divergent evolution of tissues, i.e. the divergent way of the evolutionary appearance and of histogenesis of tissues in normal development [Khlopin, 1946; Willmer, 1970] have been also described.

A major step was realization that the Modern Synthesis, the population genetic approach to evolution, does not explain macroevolution. The importance of heritable changes in development for macroevolution has been rediscovered [Goldschmidt, 1940; Waddington, 1940, 1966; Gilbert et al., 1996].

The new stage of synthesis between developmental biology and evolutionary biology was marked by the appearance of so called "evo-devo," or evolutionary developmental biology, which may be traced to the 1980s (for reviews, see [Gilbert et al., 1996; Davidson, 2006; Muller, 2007; De Robertis, 2008; Carroll, 2008]. Evolutionary developmental biology began using gene expression patterns to explain how higher taxa evolved [Gilbert et al., 1996; Goodman and Coughlin, 2000]. The remarkable homologies of homeobox genes, that control the anterior−posterior axis in vertebrates as well as in flies, and the homologous developmental pathways, including the Wnt signaling pathway, have been discovered in numerous embryonic processes (reviewed in [Gilbert et al., 1996; De Robertis, 2008]).

Several major signaling pathways, including Wnt, TGF-β, Hedgehog (Hh), receptor tyrosine kinase (RTK), nuclear receptor, Jak/STAT and Notch, control the majority of events during the development of bilaterian animals, and possibly in all metazoans. The signaling pathways act by transcriptional activation of their target genes, often using the same signal-regulated transcription factors and the same signaling pathway response elements (SPREs), as in the case of Notch, Wnt, Hh and nuclear receptor pathways [Gerhart, 1999; Barolo and Posakony, 2002; De Robertis, 2008]. Highly conserved genetic regulatory networks and regulatory sequences that play an important role in evolutionary developmental biology have been identified [Davidson, 2006].

Studies of *Drosophila*, the nematode *Caenorhabditis elegans* and zebrafish development have discovered an outstanding correlation between rates of evolution of genes expressed during different stages of ontogenesis and evolutionary morphological conservation at different stages of ontogenesis. The slowest rates of gene evolution, which are associated with stronger negative selection, were found to be associated with the so-called "phylotypic stage" of embryogenesis. The phylotypic stage is characterized by the highest morphological similarity between embryos of different species belonging to the same phylum. On the other side, the highest rates of gene evolution (associated with the positive selection) are characteristic of genes expressed in adults after they reach reproductive maturity, especially in male gonads [Domazet-Loso and Tautz, 2003; Cutter and Ward, 2005; Davis et al., 2005; Artieri et al., 2009; Garfield and Wray, 2009; Domazet-Loso and Tautz, 2010b], and in tumors, as I will discuss below. In the latter case, I mean the positive selection at the organismal level, not the somatic evolution of tumor cells.

Plant embryogenesis, which evolved independently from that of animals, is also characterized by a transcriptomic hourglass pattern, with maximally ancient and conserved transcriptomes during the phylotypic stage [Quint et al., 2012].

Correlation of phylogenetic differences at later ontogenetic stages with the expression of newly evolved genes supports the role of novel genes in the origin of evolutionary innovations and morphological novelties [Domazet-Loso and Tautz, 2003, 2010b]. These data support von Baer's third law of development (earlier developmental stages are morphologically more similar across species than later stages), the developmental constraint hypothesis (genes active in early development are involved in a larger number of downstream events and thus are more evolutionarily constrained), and Darwin's "selection opportunity" hypothesis predicting the increased divergence in adults. A constraint early/opportunity late model was suggested as best explaining the genes' divergence and their rates of evolution during ontogeny [Altieri et al., 2009; Artieri and Singh, 2010].

An important problem that evo-devo addresses is the mechanisms involved in the emergence of evolutionary innovations and morphological novelties [Muller and Newman, 2005; Muller, 2007]. The wealth of data obtained for different phyla suggests that a connection exists between progressive evolution and the origin of evolutionarily novel genes. Morphological form may also evolve by altering the expression of functionally conserved proteins. These issues are discussed in more detail in Chapter 9 of this book.

For my consideration of tumors and evolution it is important that many canonic signaling pathways discussed above are common in normal and neoplastic development, normal and tumor stem cells. On the other side, positive selection and accelerated evolution of many human tumor-related genes in primate lineage was described [Clark et al., 2003; Pavlicek et al., 2004; Nielsen et al., 2005; Crespi and Summers, 2006]. This may indicate that hereditary neoplasms, which are new to the host organism, may constitute an underlying basis for the emergence of evolutionary innovations and morphological novelties. Studying neoplasms and their gene patterns within an evo-devo framework may lead to understanding the regulatory mechanisms, gene networks and higher-order systems of organization involved in the origin of morphological and functional novelties in multicellular organisms (see Chapter 9 for further discussion of this topic).

The planktotrophic larvae of indirectly developing marine animals, which may be the illustration of the deviation mode of phylembryogenesis, incorporate so called set-aside cells that participate in adult body formation. These are cell populations with indefinite division potential and loss of cell division and differentiation controls, resembling tumors (see Parts 9.2.1, 10.9 and 10.10 for more discussion of this phenomenon).

The embryonic, fetal and infantile tumors could play the role in evolution of ontogenesis (evo-devo) (see Part 6.2 for more discussion of the possibility of this connection).

The important feature of neoplasms is their occurrence at the later stages of ontogenesis. Primary cancers usually appear only after at least one half of the life span of the affected subject has passed [Clark, 1991]. The late onset of adult cancers may be connected to the change of ontogenesis by means of the addition of final stages of development, the anaboly modus of the evolution of ontogenesis described by Severtsov [Severtsov, 1949]. If this assumption is correct, and neoplasms indeed participate in anaboly, we may look for recapitulations of some tumor features in the most recently evolved organs.

5.5 THE HUMAN BRAIN, AS THE MOST RECENTLY EVOLVED ORGAN, RECAPITULATES MANY FEATURES RESEMBLING THOSE OF TUMORS

If tumors indeed played a role in evolution of multicellular organisms, the human brain, as the most recently evolved organ, might recapitulate some features resembling those of tumors. Indeed, we can see many such features.

The brain has disproportionally increased in size in relation to other parts of the body. During development, the mammalian brain produces far more neurons than necessary. Excessive neurons are discarded in a selective way later in development, but might be used in the evolution of human brain. Some parts of the brain have evolved and enlarged as tumor-like clamps or globes, e.g. neostriatum [Rakic, 1995].

Studies of the ontogenic allometry of brain size in humans and other mammals have shown that prenatal brain-body curves for humans and macaques are identical in slope and position but humans extend the curve into postnatal ontogeny. The departure from high slope occurs just before birth in Semnopithecus, at 150 days of gestation in laboratory macaques, just after birth in chimpanzees, and not until two years after birth in humans [Count, 1947; Holt et al., 1975; Gould, 1977]. Prolongation of the high prenatal rate of brain growth into early childhood produces a human-size brain from the monkey's brain [Gould, 1977].

The proliferative layers − ventricular zone (VZ) and subventricular zone (SVZ) − enlarge during development of the central nervous system (CNS) and almost disappear in adults. The intermediate progenitor cells migrate to the embryonic SVZ. Newly generated neuron cells migrate out of the proliferative zones [Kriegstein et al., 2006].

Many of the neuron cells are aneuploid. Aneuploidy is a feature common to cancer cells [Pathak and Multani, 2006]. Chromosomal variation in neuroblasts and neurons of the developing and adult mammalian nervous system has been described. By comparison to lymphocytes, 10 times as many neuroblasts are aneuploid. The CNS, both during development and in adulthood, is a genetic mosaic, where the genetically diverse aneuploid population of cells is present concomitantly with the normal euploid cell population [Rehen et al., 2001; reviewed by Muotri and Gage, 2006].

Recombination-related genes, which may participate in the origin of somatic genomic alteration and lead to aneuploidy, are strictly required for the survival of neurons arising from the VZ [Gao et al., 1998; Chun and Schatz, 1999a, b]. Many of these genes are also connected with cancer. For example, non-homologous end-joining protein XRCC4 interplays with p53 in tumorigenesis, genomic stability and development [Gao et al., 2000].

Transposable elements L1 move in developing brain cells like they do in tumor cells, which adds to the genetic mosaicism of the CNS. Brain cells have 80 additional copies of L1 compared with other tissues. Even higher numbers of L1s are found in the hippocampus, which continues to form neurons even in adults [Muotri et al., 2005; Coufal et al., 2009; Baillie et al., 2011; reviewed in Vogel, 2011].

The level of gene expression in the brain is the highest in comparison with the other normal tissues and organs [Evtushenko et al., 1989]. As we have discussed above, tumors are also characterized by their high level of gene expression. Genome-wide analysis of cancer/testis antigen (CT) genes expression, which originally was thought to be restricted only to the testis and a variety of tumors, has led to the discovery of a group of genes with expression restricted to cancer, the testis and the brain (CTB genes). CTB genes comprise about 10% of all cancer-testis genes. For example, the *GAGE* and *MAGE* families of CT genes are expressed in the hippocampus and cerebral cortex [Hofmann et al., 2008]. Earlier study has also identified a similar group of CTB antigens [Scanlan et al., 2002b]. These findings attract attention to evolutionary connections of processes in the testis, tumors and the brain. Below we will discuss the evolutionary novelty of cancer/testis antigen genes in more detail (see Part 11.1.4).

5.5.1 The Expansion of Brain Size During Mammalian and Primate Evolution Involved Many Protooncogenes and Tumor Suppressor Genes

Reviewing the vast literature on the evolution of brain size in mammals, primates and humans suggests that many protooncogenes and tumor suppressor genes participated in the process of expansion of mammalian and especially primate and human brains.

During mammalian and especially primate evolution, the brain size has increased dramatically in relation to body size. There was an increase in the overall brain size and a disproportionate increase of the cerebral cortex during hominid evolution. The surface area of the neocortex (the most recently evolved part of the cortex) in mice, macaques and humans has an approximate ratio of 1:100:1000, respectively [Rakic, 1995; Charvet et al., 2011], without comparable increase in its thickness. This difference is connected with transformation of the *lissencephalic* cortex of rodents (adult six-layered neocortex with a smooth surface) into a highly folded *gyrencephalic* cortex characteristic of primates (adult six-layered neocortex that develops a folded surface associated with gyri and sulci). Other parts of the brain, e.g. the neostriatum, enlarged over the same evolutionary period in the form of lumps [Rakic, 1995].

Neurons of the mammalian neocortex are not generated within the cortex. The progenitors of cortical neurons are generated in the ventricular and subventricular proliferative zones (VZ and SVZ). Postmitotic cells — newly generated neurons — migrate radially to form the cortical plate. The cortical plate grows and forms a six-layered cortex. As development proceeds, VZ and SVZ become smaller and disappear in most cortical regions.

Existing hypotheses in this field suggest mechanisms explaining how this complexity could evolve [Rakic, 1995; Bond et al., 2002; Chen and Walsh, 2002; Megason and McMahon, 2002; Dorus et al., 2004; Gilbert et al., 2005; Ponting and Jackson, 2005; Kriegstein et al., 2006]. These hypotheses consider patterns of neural stem and progenitor cell divisions, cell cycle control and genes that regulate brain size during development.

A radial unit hypothesis was proposed [Rakic, 1995] according to which the cortical expansion is the result of changes in proliferation kinetics that increase the number of radial columnar units without changing the number of neurons within each unit. The ratio of symmetric and asymmetric modes of cell divisions could create an expanded cortical plate, according to this model. The size of the cerebral cortex depends on the number of contributing radial units, which in turn depends on the number of founder cells. The evolutionary difference in the size of the cortical surface between rodents and primates can be explained by the difference in the number of founder cells that is generated before the start of cortical neurogenesis [Rakic, 1995].

According to the intermediate progenitor hypothesis, neurons arise directly from radial glial cells in the VZ and indirectly from intermediate progenitor cells in the SVZ. Evolutionary adaptation that resulted in prolonged intermediate progenitor cell production would lead to the increase of the number of neurons in the cortical layers [Noctor et al., 2004; Kriegstein et al., 2006]. Development of the gyrated human neocortex is related to the expansion of intermediate progenitor cells in the outer subventricular zone (OSVZ), a proliferative region outside the ventricular epithelium [Lui et al., 2011; Molnar, 2011].

Experiments with transgenic and knockout mice and other data support the possible role of protooncogenes and tumor suppressor genes expressed in neural stem-like/progenitor cells in the evolution of brain enlargement.

The *Wnt* gene family encodes a group of conserved signaling molecules that regulate cell-to-cell interactions during embryogenesis [Nusse, 2001]. Mouse *Wnt* genes exhibit discrete domains of expression in the early embryonic CNS. At least three genes (*Wnt-3*, *Wnt-3a*, and *Wnt-7b*) exhibit sharp boundaries of expression in the forebrain that may predict subdivisions of the region later in development [Parr et al., 1993]. Homozygosity for inactivated alleles of *Wnt-1* results in the loss of midbrain and cerebellar structures in mouse embryos [McMahon and Bradley, 1990]. Ectopic expression of *Wnt-1* in transgenic mice leads to overgrowth of the neural tube [Dickinson et al., 1994]. It was shown that a mitogen gradient of dorsal midline *Wnt*s organizes growth in the CNS [Megason and McMahon, 2002].

Several components of the Wnt signaling pathway have been implicated in human tumors or experimental cancer models. *Wnt-1* was discovered as a protooncogene activated in murine mammary tumors by the mouse mammary tumor virus [Nusse and Varmus, 1982] and by now its oncogenic potential is firmly established [Nusse and Varmus, 1992].

β-catenin, a component of the adherence junctions that couple radial glia at the ventricular border, is the key downstream mediator of the Wnt signaling pathway. Inactivation of β-catenin leads to dramatic brain malformation [Brault et al., 2001]. β-catenin regulates the size of the neural precursor pool by influencing the decision to divide or differentiate. By controlling the generation of neural precursor cells, β-catenin also regulates cerebral cortical size [Chen and Walsh, 2002]. Electroporation of the dominant-active form of *β-catenin* (alone or with *Wnt-1* construct) in the chick neural tube causes the reduction of neuronal differentiation and the increase of the ventricular zone surface area [Megason and McMahon, 2002].

Activation of *β-catenin* protooncogene was shown in a variety of human cancers, including some resembling neural precursors such as medulloblastoma [Huang et al., 2000; Peifer and Polakis, 2000].

The *PTEN* tumor suppressor gene is expressed in the developing embryonic CNS in mice and humans [Luukko et al., 1999; Gimm et al., 2000]. Its germline mutations cause neurological disorders (macrocephaly) in humans [Marsh et al., 1997]. Conditional knockout mice lacking *Pten* (*Pten* is a mouse homolog of human *PTEN*) have enlarged abnormal brains due to increased cell proliferation and decreased cell death. This suggests that the PTEN tumor suppressor negatively regulates neural stem/progenitor cell proliferation [Groszer et al., 2001]. The loss of *Pten* in conditional knockout mouse models leads to an increase of the pool of self-renewing neural stem cells and to their escape from proliferation control [Groszer et al., 2006]. Similar events could participate in brain size evolution.

PTEN is frequently mutated in human cancers. It is one of the most frequently mutated genes in glioblastomas [Li et al., 1997]. $Pten^{+/-}$ mice develop germ cell, gonadostromal, thyroid and colon tumors [Di Cristofano et al., 1998].

Mutations of the *BRCA1* gene in humans [Castilla et al., 1994; Friedman et al., 1994; Futreal et al., 1994; Miki et al., 1994; Hosking et al., 1995; Merajver et al., 1995a] and the wild type allele loss [Futreal et al., 1994; Cornelius et al., 1995; Merajver et al., 1995b] are associated with predisposition to breast and ovarian cancers and manifestation of the disease. *BRCA1* is considered to be a tumor suppressor gene.

On the other hand, *Brca1* knockout mice (*Brca1* is the mouse homolog of human *BRCA1*) show profound defects in nervous system development such as failure of neural tube closure and severely retarded growth of the forebrain [Gowen et al., 1996].

The tumor-suppressor gene *ATM* plays an important role in responses to damaged DNA and neurogenesis [Andreeff et al., 2000; Lee and McKinnon, 2000; Rolig and McKinnon, 2000; Allen et al., 2001].

5.5.2 Human Cerebral Cortex as a Result of Selection for Tumor Growth

Deletion of a tissue-specific, forebrain subventricular zone (SVZ) enhancer near the tumor suppressor gene *GADD45G* (growth arrest and DNA-damage-inducible, gamma) in humans is correlated with expansion of specific brain regions [McLean et al., 2011].

Gene expression in SVZ is important because increased proliferation of SVZ intermediate progenitor cells underlies the evolutionary expansion of the neocortex in primates. In primates, the neocortex is disproportionately enlarged relative to other mammals. Cortical neurons are generated through intermediate progenitors. An evolutionary shift towards increased numbers of intermediate progenitor cells could have contributed to the increase in cortical surface area that accompanied mammalian evolution. Among mammals, the SVZ is larger in primates and especially large in humans where great cortical expansion has occurred. The human cortical SVZ has more complex histological organization and increased ratio of the size of the SVZ to VZ. It persists longer than in other mammals. All these features contribute to the expansion of the human cerebral cortex [Zecevic et al., 2005; Kriegstein et al., 2006; Charvet et al., 2011].

According to the results of Cory McLean and co-authors, species-specific loss of an ancestral SVZ tumor suppressor gene *GADD45G* enhancer provides a molecular basis for increasing production of particular neuronal cell types by regulatory changes of tumor suppressor gene in human lineage [McLean et al., 2011].

GADD45G represses cell division and activates apoptosis. Its downregulation is essential for various cancer types to escape programmed cell death [Zebrini et al., 2004]. The loss of *GADD45G* expression is described in human pituitary adenomas [Zhang et al., 2002]. The role of *GADD45G* in negative regulation of tumorigenesis suggests that, due to deletion of a tissue-specific enhancer of this gene, some tumor-like process could participate in the enlargement of SVZ and expansion of the cerebral cortex during human evolution.

So it looks like tumor suppressor genes *GADD45G*, *PTEN*, *BRCA1* and *ATM* and protooncogenes *Wnt-1* and *β-catenin* participate in CNS embryonic development, may cause brain tumors and could participate in brain size evolution by controlling neural stem-like/progenitor cells. It is possible that different capabilities of these genes are interconnected, similar to the interconnection of embryonic development, tumor formation and the evolutionary development of multicellular organisms.

5.5.3 Brain Enlargement, Microcephaly Genes and Tumors

Autosomal recessive primary microcephaly (MCPH) is a neurodevelopmental disorder. It is characterized by a small but structurally normal brain with great reduction of the cerebral cortex. There are at least seven MCPH loci; four of the genes have been identified [Woods et al., 2005].

Mutations in microcephaly genes cause a marked reduction in human brain size. That is why it was suggested that microcephaly genes participated in primate brain enlargement [Bond et al., 2002; Zhang, 2003a; Gilbert et al., 2005; Ponting and Jackson, 2005; Woods et al., 2005]. Two microcephaly genes have been well characterized: *microcephalin* (*MCPH1*) and *abnormal spindle-like microcephaly-associated (ASPM).* Both genes are expressed in VZ and SVZ during embryonic development.

5.5.3.1 MCPH1 is a Tumor Suppressor Gene Interrelated with the other Tumor Suppressor, BRCA1

Microcephalin protein contains three BRCA1 C-terminal (BRCT) domains. BRCA1 is the closest human homolog of microcephalin protein. One BRCT domain is present at the N-terminus and two are at the C-terminus of MCPH1. The BRCT domain is a phosphor-protein binding domain involved in DNA damage response and cell cycle control [Yu et al., 2003]. Depletion of microcephalin decreases both protein and transcripts of Brca1 and checkpoint kinase Chk1. So microcephalin is a DNA damage response protein involved in regulation of *BRCA1* and *CHK1*. Checkpoint defects may lead to genome instability, which predisposes organisms to cancer. Therefore, it was hypothesized that *MCPH1* is a tumor suppressor [Xu et al., 2004]. Similar function of *MCPH1* as tumor suppressor was described in studies of multiple tumor suppressor pathways negatively regulating telomerase. *MCPH1* was found by a cellular screen as a negative regulator of the telomerase protein component gene (h*TERT*) transcription. *MCPH1* thus represents a potential tumor suppressor gene [Lin et al., 2003].

The evolution of the microcephalin protein sequence is highly accelerated throughout the lineage from simian ancestors to humans and chimps. This accelerated evolution is coupled with signatures of positive selection [Evans et al., 2004b]. *MCPH1* continues to evolve adaptively in humans [Evans et al., 2005]. Most of the internal *BRCA1* sequence is also variable between primates and evolved under positive selection [Pavlicek et al., 2004]. BRCT domains in both genes are highly conserved and are themselves not subject to positive selection [Evans et al., 2004b; Pavlicek et al., 2004].

Thus *Microcephalin* and *BRCA1* genes are closely related; both contain BRCT domains and participate in DNA damage response and cell cycle control. They both are tumor suppressor genes; both are involved in brain development and subject to positive selection in the evolution of primates. Evolution of the tumor suppressor *BRCA1* locus in primates has implications for cancer predisposition [Pavliceck et al., 2004] and brain development [Evans et al., 2004b]. *Microcephalin* and *BRCA1* genes show us again that tumor and evolution processes could be interrelated.

5.5.3.2 ASPM is a Major Determinant of Human Cerebral Cortical Size, and is Overexpressed in Tumors and Testis

The *ASPM* gene is considered to be a major determinant of cerebral cortical size [Bond et al., 2002]. Mutations in the *ASPM* gene are most prevalent in people affected with MCPH. For the *ASPM* gene, similar to that for *MCPH1*, molecular phylogenetic studies suggest strong adaptive evolution in the lineage leading to humans [Zhang, 2003a; Evans et al., 2004a; Kouprina et al., 2004a; Gilbert et al., 2005]. For both *ASPM* and *MCPH1* genes, the rates of non-synonymous substitutions are accelerated in primates. Within primates, this acceleration is highest in the lineage leading to humans. For *ASPM*, the evolution accelerated more in the lineage from ape ancestors to humans. The evolutionary acceleration of *MCPH1* is more manifested between the last common ancestors of simians and the last common ancestors of the great apes. So *ASPM* might have been particularly important for human brain evolution [Gilbert et al., 2005]. Indeed, the ongoing adaptive evolution of *ASPM* in humans was demonstrated. One genetic variant of *ASPM* has been shown to arise only 5800 years ago and since then has risen to high frequency under strong positive selection [Mekel-Bobrov et al., 2005].

The *ASPM* gene is expressed in proliferating tissues, encodes for a mitotic spindle protein and is overexpressed in tumors and testis [Kouprina et al., 2005], i.e. it may belong to CTB genes discussed above.

The studies of microcephaly genes have raised a storm of discussion and criticism. It was argued that microcephaly genes are not associated with adult brain size and cognition [Dobson-Stone et al., 2007; Timpson et al., 2007; Mekel-Bobrov and Lahn, 2007; Bates et al., 2008]. But positive selection in *ASPM* is still correlated with cerebral cortex evolution across primates [Ali and Meier, 2008] and *ASPM* and *CDK5RAP2*, the other microcephaly gene, show positive associations with absolute neonatal brain size, suggesting a role in the evolution of total neuronal number [Montgomery et al., 2011]. The story of microcephaly genes will continue, but for my consideration it is important as an example of possible involvement of some tumor-like processes and tumor-related genes in the evolutionary development of the human brain.

5.5.4 Long-Term Neural Stem Cell Expansion Leads to Brain Tumors

There is evidence that higher than normal proliferation of neural stem cells (NSC) may lead to brain tumors. NSC localize in the adult SVZ. The SVZ has been considered to be the site of origin of gliomas for a long time [Smyth and Stern, 1938; Globus and Kuhlenbeck, 1942; Vick et al., 1977].

The nuclear receptor tailless (Tlx) in the adult is expressed only in NSC of the SVZ. Overexpression of Tlx in transgenic mice leads to an increase of NSC number and neurogenesis in the adult brain. It also leads to spontaneous development of glioma-like lesions and gliomas which is accelerated upon loss of the tumor suppressor *p53* gene. Tlx is overexpressed in human gliomas as well as in other types of human brain tumors. In primary human glioblastomas the *Tlx* gene is increased in copy number. These data show how NSCs of SVZ may participate in brain tumorigenesis [Liu et al., 2010]. For my consideration of evolution by tumor neofunctionalization it is especially interesting that Tlx regulates NSC expansion through the PTEN pathway discussed above in connection with brain development and evolution.

5.6 THE EUTHERIAN PLACENTA IS EVOLUTIONARY INNOVATION AND RECAPITULATES MANY TUMOR FEATURES

The other example of an organ which originated relatively recently in evolution is the eutherian placenta. Placentas have evolved independently in squamate sauropsids and in marsupial and placental mammals. Eutheria, the largest and most successful group of mammals, have the most complex and efficient placenta which belongs to the so-called chorioallantoic placenta type [Ferner and Mess, 2011].

The placenta has many similarities with tumors. It has features of local invasion and metastasis characteristic to malignant tumors. It grows in the eyes of rabbits in a manner analogous to malignant tumors after transplantation [Krebs, 1946; Gurchot et al., 1947].

The placental chorion is characterized by a high aerobic glycolysis also characteristic of tumors [Bell et al., 1928; cit. in Gurchot et al., 1947]. The DNA of the placenta demonstrates the genome-wide hypomethylation similar to that of tumors [Popp et al., 2010]. Some cancer/testis antigens are expressed in the placenta [Chen et al. 2006]. The placental trophoblast is capable of creating large ploidies that are also characteristic of many tumors [Kalejs and Erenpreisa, 2005].

Tumor markers present at high levels in the serum of patients with cancer include placental proteins (human chorionic gonadotropin (HCG), human placental lactogen, placental alkaline phosphatase, Regan isoenzyme). The increase of the HCG level takes place in a variety of nontrophoblastic neoplasms [Braunstein et al., 1973; Torosian, 1988].

So called "extravillous trophoblast" (EVT) placental cells undergo epithelial-mesenchymal transition (EMT) at the invasive front of the placenta [Vicovac and Aplin, 1996] similar to the EMT that happens at the invasive front of malignant tumors [Brabletz et al., 1998, 2005; Hlubek et al., 2007].

Tumor cells secrete proteins which stimulate tumoral angiogenesis. One such protein is vascular endothelial growth factor (VEGF), also known as vascular permeability factor (VPF) [Dvorak, 2002; Shinkaruk et al., 2003]. The placenta also produces angiogenic factors, the most important of which is VEGF [Reynolds et al., 2005].

The list of placenta/tumor similarities may be continued. The evidence that the placenta is a regulated tumor is discussed in more detail in Part 7.5.

Chapter 6

Tumors that Might Play a Role in Evolution

6.1 HEREDITARY TUMORS

In order to be evolutionarily meaningful, tumors should have the capacity to be inherited in generations of organisms. What we know about genetics of tumors suggests that it is possible. Tumor-causing genes — tumor suppressor genes, protooncogenes and DNA repair genes — exist in genomes of evolving organisms and are inherited in progeny generations, forming the basis for inheritance of certain tumors. As far as inheritance of tumors involves multi-factorial interaction of many genes and environmental factors, it is generally accepted to speak about genetic predisposition to neoplasia. Some scientists formulate this as an inherited tendency to develop tumors [Frank, 2004], or an inherited predisposition to cancer [Friedman et al., 1994]. Others use terminology like "hereditary cancers" and "inherited cancers" [Knudson, 1996; Kinzler and Vogelstein, 1997; Shakya et al., 2011]. I will use the latter terminology.

Since Alfred Knudson formulated his two-hit hypothesis explaining the origin of childhood retinoblastoma, the importance of germline mutations in the genetic predisposition to neoplasms has been recognized. It was hypothesized that in the inherited tumors, one mutation is inherited through the germinal cells and the second occurs in somatic cells [Knudson, 1971]. Knudson also suggested that childhood tumors may represent the consequence of the homozygous state of those genes which are involved in differentiation of embryonal tissues [Knudson, 1979], and that the normal allele may be regarded as anti-oncogenic [Knudson, 1989]. Thus one of the first inferences about tumor suppressor genes has been established.

There are other examples of germline mutations in tumor suppressor genes, which cause hereditary cancers. Germline mutations in the breast cancer susceptibility gene *BRCA1* increase the lifetime risk of breast cancer among female mutation carriers to 82% [King et al., 2003]. The estimate is that 88% of breast-ovarian cancer families are due to mutations in *BRCA1* [Easton et al., 1993; Narod et al., 1995]. Many mutations in the *BRCA1* gene have been reported in families with a predisposition to both inherited and

Evolution by Tumor Neofunctionalization.

sporadic breast and ovarian cancers [Castilla et al., 1994; Friedman et al., 1994; Futreal et al., 1994; Miki et al., 1994; Hosking et al., 1995; Merajver et al., 1995a,b]. Evolution of the tumor suppressor *BRCA1* locus in primates is connected with cancer predisposition [Pavlicek et al., 2004].

Nearly all carriers of the *APC* gene mutations develop adenomatous polyposis and colon cancer, and mutations in DNA mismatch repair genes cause hereditary nonpolyposis colon cancer. Germline mutations in the *p53* gene are connected with Li-Fraumeni syndrome and elevated risk for soft-tissue sarcomas, osteosarcomas, brain tumors, breast cancers and leukemias [Fearon and Vogelstein, 2000].

Altogether, almost 50 different inherited cancers or cancer syndromes are described in humans [Knudson, 1989]. Germline mutations of *PTEN* and *LKB1* tumor suppressor genes cause Cowden disease (CD) and Peutz-Jeghers syndrome (PJS), respectively [Smith and Ashworth, 1998]. CD and PJS are autosomally dominantly inherited diseases that share clinical features, such as multiple tumors. *PTEN* and *LKB1* tumor suppressor genes display overlapping expression patterns during embryonic development [Luukko et al., 1999].

For my consideration of the possible role of tumors in evolution, it is interesting that germline mutations in tumor suppressor *PTEN* are involved in hamartomatous polyposis syndromes [Marsh et al., 1997; Smith and Ashworth, 1998]. Hamartomas are benign, hyperplastic, disorganized masses of cells and tissues that may play a role in evolution, according to my hypothesis of evolution by tumor neofunctionalization.

Genetic predisposition to neoplasia occurs in fishes. For example, the cancer-prone Atlantic tomcod (*Microgadus tomcod*) is characterized by an increased prevalence of hepatic neoplasms that is associated with a polymorphism in the cytochrome P4501A1 gene [Roy et al., 1995]. Hereditary tumors are frequent in hybrid fishes, such as goldfish × carp hybrids, platyfish × swordtail hybrids, and various inbred strains or varieties of fishes such as Japanese medaka, goldfish and ornamental koi carp (reviewed in [Groff, 2004]).

The high incidence of tumors in inbred strains of laboratory animals, especially in mice and rats, is a well-known phenomenon. Some strains of mice have a 90−100% incidence of certain tumors by a certain age [Huxley, 1958; Benirschke et al., 1978]. Similar populations of tumor-bearing animals could have played a role in evolution, even though the proportion of hereditary tumors may be smaller as compared with non-hereditary ("sporadic") tumors (see Part 10.6 for more discussion of the possible role of populations of tumor-bearing organisms in evolution).

6.2 FETAL, NEONATAL AND INFANTILE TUMORS

Embryonic tumors are described in humans and animals [Willis, 1953]. The existence of embryonic tumors supports the theory of F. Durante and J. Cohnheim which I discussed above in Part 5.1. Wilm's tumor and neuroblastomas of infants are the examples. Wilm's tumors originate from embryonal cells which differentiate with age. Neuroblastomas are composed of undifferentiated embryonal neural cells which can differentiate into benign mature cells in older children [Sell, 2004].

Neonatal tumors generally are less malignant and more often differentiate or regress than do tumors of adults. They are characterized by benign clinical behavior despite a malignant histological picture. Neonatal malignant Wilm's tumors probably do not exist. Neuroblastomas show a high cure rate during the first year of life and a high rate of spontaneous regression [Bolande, 1971; Meizner, 2000, 2011].

The majority of spontaneously regressing neuroblastoma cases originally described by T.C. Everson and W.H. Cole were detected in infants [Everson and Cole, 1966]. The incidence of *in situ* neuroblastomas determined in autopsies of infants dying of other causes is much higher than the incidence of manifested neuroblastomas. This means that a large proportion of *in situ* neuroblastomas must undergo spontaneous regression [Beckwith and Perrin, 1963]. More recently, Japanese infant screening programs for neuroblastoma have shown increasing rates of incidence due to early-stage neuroblastomas while more advanced tumors and deaths due to neuroblastoma remained at the same level [Yamamoto et al., 1995; Ajiki et al., 1998; Honjo et al., 2003]. It was hypothesized that tumors identified by screening spontaneously regress and fundamentally differ from the symptomatic tumors [Yamamoto et al., 1995; Honjo et al., 2003; Kramer and Croswell, 2009].

Other authors came to similar conclusions much earlier. H.G. Wells wrote in 1940 that congenital neoplasia is fundamentally different from the common cancers and true malignancy is rare at the early stages of development [Wells, 1940]. Based on the benignity of neonatal tumors, R.P. Bolande formulated the concept of cancer repression in early life. The concept suggests that a certain grace period may exist during fetal life and early infancy during which the organism is relatively resistant to the development or progression of malignant disease due to specific repressive influences on the cancer genome [Bolande, 1971]. This may be important for the evolution of ontogenesis by the addition of final stages, i.e. anaboly.

The latest data support the fundamental difference of childhood tumors. Studies of the genetic landscape of the childhood cancer medulloblastoma revealed a major genetic difference between adult and childhood solid tumors, i.e. the smaller number of genetic alterations observed in childhood tumors. Other important differences include much more frequent mutations of genes involved in normal development, such as *Hedgehog* and *Wnt* pathway genes, and inactivating mutations of the histone-lysine N-methyltransferase genes *MLL2* or *MLL3*. This raises the question of whether alterations of developmental pathways, including heritable epigenetic alterations, are responsible for childhood tumors [Parsons et al., 2011].

The peculiarities of congenital, fetal and infantile tumors could play a role in the evolution of ontogenesis (evo-devo) through evolutionary changes of embryonic anlages and the addition of final stages of ontogenesis (the anaboly mode of phylembryogenesis).

6.3 BENIGN TUMORS, CARCINOMAS *IN SITU* AND PSEUDODISEASES

Rupert A. Willis, in his classic book *Pathology of Tumors*, described benign, or "innocent," tumors in the following way: the structure of benign tumors is often typical of the particular tissue or organ; growth usually is purely expansive, and a capsule is formed; the rate of growth usually is slow; mitotic figures are scanty; the benign tumor may come to a standstill or regress; metastasis is absent; the tumor may be dangerous only because of position, or accidental complications, or production of excess of hormone. On the other hand, the structure of malignant tumors is often atypical, the differentiation is imperfect, and the most anaplastic tumors are the most malignant; tumor growth is infiltrative as well as expansive; the strict encapsulation is absent; tumor growth may be rapid, with many mitotic figures; tumor growth rarely ceases and usually is progressive to a fatal termination; metastasis is frequently present; malignant tumors are intrinsically dangerous because of progressive infiltrative growth and metastasis. Willis also describes borderline tumors which occupy the position between benign and malignant tumors, such as the papillary growth of the bladder with a complete range from slowly growing papilloma to rapidly growing infiltrative papillary carcinoma [Willis, 1953].

From these descriptions, which have not changed in past years (see also discussion of the structure of benign and malignant tumors in Part 4.4), it is evident that benign and some borderline tumors are more likely candidates for playing a role in the evolution of multicellular organisms than are malignant tumors. Malignant tumors at the late stages of progression cannot play such a role, although cases of spontaneous regression of malignant tumors are described and the placenta, a tumor-like organ in eutherians, has invasive potential.

Benign tumors are widespread in nature. As described above in Part 3.1, the tumors of fishes are mostly benign. The number of benign neoplasms in wild [Lombard and Witte, 1959] and domestic animals [Priester and McKay, 1980] is comparable to the number of malignant neoplasms. In humans, special studies have shown comparable or even higher prevalence of benign tumors versus malignant tumors [Neugut et al., 1997; Yu et al., 2007; Kirimoto et al., 2008; Stravodimos et al., 2008; Shishegar et al., 2011; Russo et al., 2012; Toshida et al., 2012]. Some benign tumors occur in a considerable proportion of the human population, e.g. colorectal adenomas [Neugut et al., 1997], benign prostatic hyperplasia in men [Bostwick et al., 1992] and fibroadenoma of the breast in women [Marchant, 2002; El-Wakeel and Umpleby, 2003; Guray and Sahin, 2006].

Carcinoma *in situ* is a tumor lesion that does not pass the barrier of the basement membrane and does not invade underlying tissues [Schauenstein, 1908; Rubin, 1910, 1918; Broders, 1932; Fennell, 1955; Fennell and Castleman, 1955]. Two sources of data suggest that carcinomas *in situ* may belong to so called pseudodiseases, i.e. tumor lesions that have no malignant potential despite histological diagnosis of cancer. The first is autopsy studies which show that a considerable proportion of people who had died from other causes had carcinomas *in situ*, or histologically confirmed cancers, or other occult tumor lesions [Breslow et al., 1977; Harach et al., 1985; Welch and Black, 1997; Haas et al., 2008]. The other is the results of the early cancer diagnosis programs. These results clearly show the rise in incidence of localized tumor lesions, and more stable trends in incidence of distant lesions or advanced diseases and mortality rates. The implication is made that early screening detects large numbers of cancers that never metastasize [Kramer, 2004; Kramer and Croswell, 2009]. The authors theorize that the natural history of asymptomatic tumor lesions detected by early screening is essentially unknown. They suggest that these lesions are less aggressive and more slow-growing, and that the majority of individuals with such occult disease will never develop the target cancer. They call such lesions the "pseudodisease" [Kramer, 2004; Kramer and Croswell, 2009] (compare with "pseudoneoplasms," discussed in Part 3.1).

A similar conclusion, that infantile tumors identified by screening differ fundamentally from the symptomatic tumors [Yamamoto et al., 1995; Honjo et al., 2003], was discussed above in Part 6.2.

Many of these studies have been performed for clinical purposes. But for my consideration of the possible evolutionary role of tumors it is important that in populations of highly organized multicellular organisms there is a considerable proportion of organisms bearing tumors that do not kill these organisms and may be used as extra cell masses for further progressive evolution.

6.4 TUMORS AT THE EARLY AND INTERMEDIATE STAGES OF PROGRESSION

Tumor progression is a stepwise acquisition of permanent new characters during tumor development leading to the increase in tumor malignancy. It was first described by P. Rous and J.W. Beard for the progression to carcinoma of virus-induced papillomas [Rous and Beard, 1935] and H.S.N. Green for spontaneous mammary cancer [Green, 1940] in rabbits. Leslie Foulds, in a series of articles (reviewed in [Foulds, 1954]), formulated several major principles of tumor progression, including the principle of independent progression of characters and the principle of different paths of progression leading to similar end-points. He proposed that "the structure and behavior of tumors are determined by numerous 'unit characters' which within wide limits are independently variable and capable of independent progression and which can be assorted and combined in a variety of ways... A tumor may have one or more of the recognized criteria of malignancy; it may have all of them and kill the patient..." [Foulds, 1954]. The important circumstance is that each unit character is variable within wide limits and may progress to greater intensity. Progression can bring a tumor to malignancy, to regression or to any intermediate stage [Foulds, 1954].

The "unit characters" of Foulds well correspond to the "hallmarks of cancer" defined as a certain number of molecular, biochemical and cellular traits shared by most and perhaps all types of cancer and acquired during the multistep development of tumors [Hanahan and Weinberg, 2000, 2011]. The hallmarks of cancer, according to Douglas Hanahan and Robert Weinberg, include sustaining proliferative signaling, evading growth suppressors and differentiation, resisting cell death, enabling replicative immortality, inducing angiogenesis, activating invasion and metastasis, reprogramming of energy metabolism and evading immune destruction [Hanahan and Weinberg, 2000, 2011]. Hanahan and Weinberg also formulated the concept of alternative or parallel pathways of tumorigenesis, which states that the hallmarks may appear at different times and in different sequence in different progressions, but the biological endpoint that is ultimately reached − cancer − should possess the combination of all hallmarks [Hanahan and Weinberg, 2000]. This concept is in good correspondence with the Foulds principle of different paths of progression leading to similar end-points.

Focal hyperplasia and benign tumors, carcinomas *in situ*, primary invasive cancers and metastatic cancers are usually thought of as the major successive steps in tumor progression, and appropriate classification of the sequential lesions of tumor progression was suggested [Clark, 1991]. A similar continuum but with a somewhat different formulation would include hyperplasia, metaplasia, atypical metaplasia, dysplasia, carcinoma *in situ*, invasive cancer and anaplastic cancer [Vincent, 1985]. Here metaplasia means the substitution of one type of differentiated adult cell by another, dysplasia means atypical cells even more different from the norm and from each other and anaplasia implies cells that have no differentiated properties of their tissues of origin [Vincent, 1985]. The latter description of the major stages of tumor progression emphasizes the loss of differentiation traits during tumor progression (due to deviation of stem cells from the normal differentiation course, not because of dedifferentiation of mature cells) which is important for my consideration of the possible evolutionary role of tumors.

An interruption in disease progression often occurs, and in many organs preneoplastic lesions are considerably more prevalent than aggressive cancers and never develop into malignant tumors (see discussion above in Part 6.3).

To quote W.H. Clark, "Viewed as a process, tumor progression is not obligatory. Indeed, the net directionality of lesions early in a tumor progression system is toward regression" [Clark, 1991].

The dormancy of metastatic cells and micrometastases, which never progress to macroscopic metastatic tumors after successful dissemination [Townson and Chambers, 2006; Aguirre-Ghiso, 2007; Strauss and Thomas, 2010], and spontaneous regression of malignant tumors (see Parts 6.5 and 6.6) were also described.

Foulds considered the nature of progression as similar to epigenetic development in embryology. He thought that progression resembles embryonic differentiation and that similar biologic laws operate in normal and neoplastic development [Foulds, 1954]. He used the "canalization" and "epigenetic landscape" terminology of Waddington [Waddington, 1940]. But after the papers of John Cairns and Peter Nowell were published [Cairns, 1975; Nowell, 1976], the predominant paradigm for many years became the concept of genomic instability and clonal selection of tumor cells as the general mechanisms of tumor progression [Nowell, 1986; Hill, 1990; Lengauer et al., 1998; Morange, 2012; Wu et al., 2012]. Nevertheless, the last decade was marked by the revival of epigenetic thought and the aberrant epigenetic landscape in cancer is being rediscovered [Feinberg et al., 2005; Esteller, 2007; Berdasco and Esteller, 2010].

Between six and ten successions of genetic and epigenetic changes are necessary to generate malignant cancer cells [Bernards and Weinberg, 2002]. According to other estimates, three to twelve mutations are necessary for development of different tumors [Renan, 1993].

From the point of view of my hypothesis of the possible evolutionary role of tumors, primary tumors at the earlier stages of progression, or tumors with intermediate progression of "hallmark" characters and/or an incomplete set of those characters, or the dormant metastatic tumors may provide the extra cellular material for the evolution of organisms. For example, selection of the early stages of papillomatosis produced normal structures — symbiovilli — in voles (see Part 7.4, below).

6.5 TUMORS THAT SPONTANEOUSLY REGRESS

Seasonal tumors with complete regression have been described in fish and newts [Pfeiffer et al., 1979; Asashima et al., 1982, 1985; Getchell et al., 1998; Holzschu et al., 2003], a phenomenon not known in mammals. The spontaneous regression of tumors is common in mature laboratory rodents, for example carcinogen-induced skin papillomas in mice [Foulds, 1954; Anisimov et al., 2005]. So-called "responsive" spontaneous mammary tumors in mice grow during pregnancy, reach a peak before parturition and regress after that [Foulds, 1954]. The genetic control of spontaneous regression of erythroleukemia was studied in parental and hybrid mice in which leukemia was induced by the regressing strain of the Friend virus. The data suggested that regression is influenced by several genes shown to affect recovery from leukemia in other systems [Dietz et al., 1981].

Spontaneous regression or remission as the partial or complete disappearance of an untreated (or inadequately treated) malignant tumor is well documented in humans, although it is considered to happen very rarely [Everson and Cole, 1956, 1966; Everson, 1964; Boyd, 1966; Cole, 1974; Evans et al., 1976; Challis and Stam, 1990]. Spontaneous regression was most commonly noted in neuroblastoma, hypernephroma, choriocarcinoma and malignant melanoma [Everson, 1964]. Everson and Cole demonstrated that over 63% ($n = 111$) of cases of spontaneous regression were accounted for by five types of cancer: kidney, neuroblastoma, malignant melanoma, choriocarcinoma and bladder [Everson and Cole, 1966]. If the cases of retinoblastoma, lymphoma, leukemia and breast cancer were added, than these nine cancers accounted for 69% ($n = 514$) of all cases of spontaneous regression. In total, 741 individuals presented spontaneous regression of cancer between 1900 and 1987 [Boyd, 1966; Everson and Cole, 1966; Challis and Stam, 1990]. New cases of spontaneous regression of malignant tumors in humans continue to be registered nowadays and the phenomenon of spontaneous regression continues to attract the attention of numerous oncologists [Maurer and Koelmel, 1998; Abdelrazeg, 2007; Bir et al., 2009; Salim et al., 2009; Herreros-Villanueva et al., 2012]. According to Uwe Hobohm, about 1000 cases of spontaneous regression have been described in medical literature during the past century [Hobohm, 2009].

It was noticed that many cases of spontaneous regression of tumors have been linked to acute bacterial infections and fever. In different studies, 25 to 80% of spontaneous regressions were connected with prior fever. William Coley was the first to systematically study the influence of bacterial infection, bacterial products (Coley's toxins) and fever induced by them on malignant tumors in humans. Coley discovered that such treatment is more powerful in sarcoma than in carcinoma. In treating primary inoperable sarcoma, a cure rate higher than 10% (up to 80%) was accomplished by using this pioneering approach [Coley, 1893, 1894; Nauts et al., 1953; Wiemann and Starnes, 1994; Hobohm, 2009]. Bacterial immunotherapy of tumors is being seriously considered by oncologists armed with modern immunologic knowledge. Transferring of this approach into current oncological clinics is on its way [Linnebacher et al., 2012].

The counterpart of Coley's toxin effect in laboratory animals is known as the "hemorrhagic necrosis" of transplanted tumors caused by material from Gram-negative bacteria. Studies of the hemorrhagic necrosis produced by endotoxin have led to the discovery of an endotoxin-induced serum factor that causes necrosis of tumors (tumor necrosis factor, TNF) [Carswell et al., 1975]. This discovery, in its turn, has led to the current clinical trials of recombinant cytokines as antitumor agents, and to the formulation of an immunological hypothesis which states that spontaneous regression of tumors is the result of immune attack [Wiemann and Starnes, 1994; Hobohm, 2001, 2005, 2009]. It is of interest that only immunogenic tumors, which are predominantly of mesodermal origin, can be cured by endotoxin-related therapy. In connection with this circumstance, B. Wiemann and C.O. Starnes thoughtfully wrote in their 1994 review, "It may well be that the treatment that Coley developed was taking advantage of a system that was designed for other purposes, i.e. control over embryonic development and differentiation, rather than immune surveillance of cancer" [Wiemann and Starnes, 1994].

Indeed, some authors have pointed out that normalization of tumors *in vivo* is associated with differentiation [Evans et al., 1976; Shvemberger, 1986]. For example, spontaneous regression of neuroblastoma usually consists of complete disappearance of the tumor, but in some cases, maturation to benign ganglioneuroma occurs [Evans et al., 1976]. Similar processes could participate in evolution of organisms by tumor neofunctionalization.

6.6 SUSTAINABLE TUMOR MASSES

Sustaining a stable tumor mass may be a strategy in cancer therapy and in the evolution of tumor-bearing organisms. As discussed by Robert Gatenby, there are similarities between cancer cells and invasive species that make it difficult to eradicate both of them [Gatenby, 2009]. For example, invasive insect species develop resistance to insecticides. Cancer cells develop resistance to cytotoxic drugs. Invasive species are currently successfully controlled by strategies of "integrated pest management" that restrict population growth but do not completely eradicate infestation. Lower-dose chemotherapy that aims to maintain a stable tumor volume ("adaptive therapy" or "treatment-for-stability") leads to longer (or indefinite) survival of tumor-bearing animals. Prolonged stable tumor volume can be achieved [Gatenby, 2009; Gatenby et al., 2009].

This is similar to what might happen in evolution with tumor-bearing organisms. Through a variety of feedback and network interactions, including immunological interactions, the extra tumor cell mass could be stabilized and sustained for a certain period of time until it differentiates and gains a function in the organism due to expression of novel genes.

Indeed, immunologically mediated tumor dormancy and equilibrium with the host has been shown experimentally. The immune system can control the outgrowth of occult primary carcinomas and metastases for extended periods of time. It was suggested that equilibrium may extend throughout the life of the host [Schreiber et al., 2011].

Chapter 7

Tumors that have Played a Role in Evolution

7.1 THE NITROGEN-FIXING ROOT NODULES OF LEGUMES

The nitrogen-fixing root nodules of legumes symbiotically associated with rhizobia are an example of the situation in which tumors acquire a new function in the plant.

Leghemoglobin, which is necessary for nitrogen fixation and makes about 40% of the soluble protein in root nodules [Beringer et al., 1979], is encoded by the host plant [Anderson et al., 1996; Hardison, 1996; Kundu et al., 2003]. The leghemoglobin gene originated as a result of duplication of an ancestor gene, which encoded so-called non-symbiotic hemoglobin, and which then diverged in the course of legume evolution [Anderson et al., 1996; Hardison, 1996; Kundu et al., 2003]. At the time of its origin the leghemoglobin gene was a novel gene for legumes. The enzyme nitrogenase, which reduces atmospheric nitrogen, is encoded by the bacterial genome [Beringer et al., 1979] and was also new for the host plant. Nitrogen fixation was an important new function that involved many genes novel for the plant. That is why extra cell mass – the root nodule – was necessary to support the expression of all these genes and to provide material for the origin of new cell types. After the novel function of nitrogen fixation was fully developed and regulatory feedback loops originated, root nodules became new organs of the legume plants.

The important difference of nitrogen-fixing root nodules from the tumors induced by *A. tumefaciens* – crown galls – is that nitrogen fixed in the nodules is used by the host plant, resulting in the emergence of a regulated function and the novel organ. By contrast, opines generated in tumors induced by *A. tumefaciens* are used exclusively by the bacteria, so crown galls are nothing but parasitism, which is harmful for the host plant. That is why crown galls resemble true tumors of animals.

The genus *Rhizobium* and the genus *Agrobacterium* belong to the same family Rhizobiaceae. Studies carried out in recent decades have shown that *A. tumefaciens* transforms plant cells with a single-stranded DNA copy

(T-DNA: transferred DNA) of a discrete region of its large tumor-inducing plasmid (Ti plasmid) [Chilton et al., 1977; Zupan et al., 2000]. T-DNA includes oncogenes coding for auxins and plant cytokines, which induce cell division and tumor development. Besides those, T-DNA includes the gene coding for an enzyme involved in the synthesis of opines, which are amino acid derivatives used by bacteria (but not their plant host) as carbon, nitrogen and energy sources. Severely affected plants become sick, undersized and unproductive. Thus, crown galls provide an example of true infectious tumor pathology in plants.

Root nodules in plants where the nitrogen fixation function is destroyed due to mutations also acquire similarities with tumors and become harmful to the host plants [N.A. Provorov, personal communication]. This confirms that the origin of the new function was essential for the appearance of the novel plant organs (root nodules) from tumors.

7.2 MELANOMATOUS CELLS AND MACROMELANOPHORES OF *XIPHOPHORUS* FISHES

Another example of a tumor role realized in evolution may be found in fishes of the genus *Xiphophorus*, which feature giant cells, macromelano-phores, a novel cell type localized laterally along their bodies and playing an important role in camouflage and ornamentation. Classic experiments of M. Gordon [Gordon, 1927] and C. Kosswig [Kosswig, 1928] showed that, in the progeny obtained by backcrossing of *Xiphophorus maculatus* and *Xiphophorus helleri*, some fishes develop melanomae. A genetic model was suggested to explain these and similar hybridization experiments [Ahuja and Anders, 1976; Anders, 1991]. The suggested simple two-locus model implies that *X. maculatus* carries a "tumor gene," *Tu*, which is mapped to the sex chromosome, and an autosomal regulatory locus *R*, which suppresses the activity of the tumor locus. *X. helleri* has neither *Tu* nor *R*. When *R*-containing autosomes are eliminated by crossing, the *Tu* locus is released from control, resulting in its enhanced expression and enlarged spots on the bodies of *R*-heterozygous fishes. The absence of both copies of *R* in *Tu*-containing backcross hybrids results in malignant melanoma.

I suggested that the opposite occurred in evolution. First, a population of fishes with a high predisposition to tumor development could appear because of the emergence of the locus *Tu*. Thereafter, the gene *R* could originate to block melanoma development, eventually resulting in the emergence of macromelanophores. A fish population previously highly predisposed to melanoma would thus evolve into a population with a novel cell type [Kozlov 2008, 2010].

The *Tu* locus was found to include two genes: *Mdl*, which determines the different phenotypes of the macromelanophore (pigmented) pattern, and *Xmrk*, which is an oncogene *(Xiphophorus* melanoma receptor kinase).

Both genes are relatively new in evolution and emerged shortly before the emergence of the genus *Xiphophorus* (5—6 million years ago) [Weis and Schartl, 1998].

It was shown that *Xiphophorus* females prefer males with the spotted caudal (Sc) melanin pattern, which is associated with the presence of the *Xmrk* oncogene and serves as the site of melanoma formation. Sexual selection appears to be maintaining this functional oncogene because of a mating preference for Sc, as well as the exaggeration of this male trait [Fernandez and Morris, 2008].

Macromelanophores became advantageous because they acquired a function, which was fixed by natural selection. Thus, macromelanophores in *X. maculatus* may be nothing but melanomatous cells which were stabilized by the fortunate combination of the respective oncogene and suppressor gene with the emergence of the function associated with the *Mdl* gene.

7.3 THE HOOD OF GOLDFISHES, AN ARTIFICIALLY SELECTED BENIGN TUMOR

Varieties of goldfish which develop head growths, or hoods (Oranda and Red cap Oranda, Lionhead, Ranchu) have been bred by selectionists over several hundred years. In this case, the fancy look of hoods on the heads of the goldfishes determined their selection and the successful growth of the population of fish endowed with this trait. According to our analysis [Kozlov et al., 2012] these hoods are benign papillomas. That is, benign tumors were artificially selected for. To our knowledge, this is the first example of artificial selection of benign tumors described in the literature. See more discussion of this example in Part 11.2.

7.4 MALIGNANT PAPILLOMATOSIS AND SYMBIOVILLI IN THE STOMACHS OF VOLES

I will now present an example of malignant neoplasm that may have played a role in the evolution of rodents. Hereditary malignant neoplasm giving rise to the formation of multiple macrovilli in the cardiac portion of the stomach have been found in *Microtus abbreviatus* (Microtinae, Cricetidae), a vole endemic to St. Matthew Island in the Bering Sea [Rausch and Rausch, 1968]. Neoplasms leading to papillomatosis of stomach corneal epithelium were also described in the pine mouse (*Microtus (Pitimus) pinetorum*) [Cosgrove and O'Farrell, 1965], the lemming and the house mouse, although in most rodents papillomatosis is rare [Vorontsov, 2003]. Structures with almost the same morphology were found in the stomach of laboratory mice treated with carcinogens [Stewart, 1941]. Papillomatosis leads to death in adult and old voles, but voles affected with papillomatosis still survive long enough to reproduce.

On the other hand, in several rodent genera the cardiac portion of the stomach is lined with normal macrovilli. These genera are *Mystromis* (Cricetinae, Cricetidae), *Myospalax* (Myospalacinae, Cricetidae), *Tachyoryctes* (Tachyoryctinae, Cricetidae), and *Cryptomys* (Bathyergidae). Macrovilli favor the development of symbiotic flora and for this reason they are called symbiovilli. Growth of the corneal epithelium of the cardiac portion of the stomach serves as a morphological basis of symbiovilli [Vorontsov, 2003].

The morphological and histological patterns of macrovilli, which developed as neoplastic growth in the voles *M. abbreviatus*, are very similar to those of the symbiovilli found in *Myospalax*, *Tachyoryctes*, and *Cryptomys*. Symbiovilli represent a normal character developed independently in four genera of rodents. Three of them (*Myospalax*, *Tachyoryctes*, and *Cryptomys*) are burrowing rodents.

Nikolai Vorontsov, the author who described this interesting case, interpreted it as fixation of Goldschmidt's macromutation as a species and genus character. He suggested that at the early stages of papillomatosis the pathogenic morphogenesis creates favorable conditions for the development of symbiotic microflora, which gives a selective advantage to the affected animals. Thus the malignant neoplasm at the early stage of progression serves as a preadaptation for the growth of symbiotic flora in the stomach [Vorontsov, 2003]. This malignant neoplasm is normalized due to emerging functional feedbacks and produces normal structures of the stomach, I would add.

Papillomatosis and symbiovilli are examples of how tumors at the early stage of progression could be used for the origin of normal functional structures, as I discussed above in Part 6.4.

7.5 EUTHERIAN PLACENTA, THE REGULATED TUMOR

The placenta is a tumor-like structure which has features of local invasion and metastasis characteristic to malignant tumors. Placental trophoblast is the only normal tissue that metastasizes but whose growth is restricted by the host [Krebs, 1946].

Placentas are classified as hemochorial, endotheliochorial or epitheliochorial. A most invasive hemochorial placenta may represent the ancient condition of Eutheria [Wildman et al., 2006; Ferner and Mess, 2011].

One of the differentiated structures of the placenta is the multinucleated syncytiotrophoblast which is formed by the fusion of cytotrophoblast cells. It is known that integrated retroviruses can cause cell fusion, which is connected with viral envelope glycoprotein incorporated into the cellular plasma membrane. The presence of endogenous retroviral particles was demonstrated in human and animal placentas, and in some tumors. This led J.R. Harris to hypothesize that ancient retroviral infection gave rise to placentas in the ancestors of mammals some 200 million years ago. To quote Harris, "I would therefore like to propose the hypothesis that at an early stage of placental mammals, developing embryos became infected at an early intrauterine stage with a retrovirus, which gave rise to cellular proliferation and creation of the trophoblast. This could have led to the formation of the highly invasive 'tumor-like' vacuolated and microvillated syncytial plate and a primitive placenta. Some of the more slowly dividing truly embryo-forming cells must also have contained the retroviral progene, thereby retaining this genetic information in the germ line cells, so that future generations of embryos would continue to developmentally express the trophoblast and its syncytial plate as a retained cellular feature during early embryonic growth" [Harris, 1991].

Indeed, *syncytin*, a domesticated retroviral gene that plays a role in placental biology, has been identified in humans. This is the envelope gene of human defective endogenous retrovirus HERV-W. It is expressed in multinucleated placental syncytiotrophoblasts and may mediate placental cytotrophoblast fusion [Blond et al., 1999; Mi et al., 2000]. Genome-wide screening for fusogenic human endogenous retrovirus envelopes later identified the second syncytin gene, *syncytin 2*, which was conserved in primate evolution [Blaise et al., 2003]. Two syncytin genes of retroviral origin, *syncytin-A* and *syncytin-B*, homologous but not orthologous to the human syncytin genes, were also discovered in rodents [Dupressoir et al., 2005] and one, *syncytin-Ory 1*, in the rabbit [Heidmann et al., 2009]. Murine *syncytin-A* demonstrated the critical role in placentation in knockout mice [Dupressoir et al., 2009], and *syncytin-Ory 1* is expressed in rabbit placenta and demonstrated fusogenic activity [Heidmann, 2009]. *Syncytin-1* and *syncytin-2* genes entered the primate lineage 25 to 40 million years (Myr) ago, *syncytin-A* and *syncytin-B* entered the rodent lineage 20 Myr ago, and the *syncytin-Ory 1* gene entered the lagomorpha order 12−30 Myr ago. The identification of syncytin genes in three different orders of mammals suggests that independent retroviral infections have led to recurrent origin of syncytiotrophoblast-containing hemochorial placentas in the course of evolution [Heidmann et al., 2009]. (See also Part 8.4 for more discussion of *syncytin* and domesticated *gag* genes).

In the eyes of rabbits, transplanted human trophoblast grows in a manner analogous to malignant tumors. Thus heterologous transplantability as a criterion of malignancy is fulfilled by trophoblast [Gurchot et al., 1947]. The malignancy of trophoblast was also demonstrated in culture, where it completely destroyed the embryonic tissue [Maximov, 1924; cit. in Gurchot et al., 1947]. The malignancy of trophoblast and its morphological features resemble those of chorionepithelioma, but its overgrowth is prevented by factors produced by both the embryo and the mother [Maximov, 1924; Krebs, 1946; Gurchot et al., 1947]. Charles Gurchot and co-authors described placental trophoblast as potentially "violently malignant" [Gurchot et al., 1947].

The question of "what makes the human placenta different from a tumor" [Lala et al., 2002] arises. Molecular mechanisms that extravillous trophoblast (EVT) cells use to invade the uterine deciduas are the same as those in cancer cells, including binding to extracellular matrix (ECM) components, degradation of ECM by production matrix metalloproteases (MMPs) and migration through the degraded ECM which requires the presence of $\alpha_5\beta_1$ integrin. But normal EVT cell lines are not immortal and do not demonstrate anchorage-independent growth or tumor formation in nude mice. EVT cells *in situ* are under strong control of many growth factors, growth factor-binding proteins, proteoglycans and ECM components. Transforming growth factor (TGF)-β produced by maternal decidua is the major negative regulator of EVT cell proliferation, migration and invasiveness. Premalignant and malignant EVT cell lines are TGF-β resistant. The loss of receptor-activated Smad3 may be the reason for disruption of TGF-β signaling in choriocarcinomas [Lala et al., 2002].

Thus the placenta is a tumor-like organ, from the standpoint of the discussion in Part 4.4. Its origin could be connected with retroviral infection in ancestor organisms which induced tumor growth and syncytiotrophoblast formation. The direct contact of invasive trophoblast with maternal blood turned out to be highly adaptive for the favorable development of mammalian embryos, was supported by natural selection and led to evolution of the invasive hemochorial placenta. The powerful regulation of invasive placental trophoblast by TGF-β is a very important example of how malignant properties are efficiently inhibited by functional negative feedback from the host organism. I will use this example in my discussion of the mechanisms of tumor neofunctionalization (see Part 10.4).

Syncytin genes and *gag*-derived domesticated genes are essential for placenta formation (see [Volff, 2006] for review). They were evolutionarily new for the ancestral host organism when the retroviral infection, which led to the placenta's origin, happened. These evolutionarily novel genes were expressed in tumors induced by retroviruses and participated in the origin of placental function and the evolution of the new tumor-like organ. Other evolutionarily novel genes are also expressed in the placenta [Knox and Baker, 2008].

I think that the expression of evolutionarily novel genes in tumors that leads to tumor neofunctionalization, i.e. acquisition of the new organismal function by tumors, is a general phenomenon in the evolution of multicellular organisms.

7.6 THE "EVOLUTION VS. PATHOLOGY PARADOX" MAY ALSO EXIST FOR TUMORS

As already discussed in Part 2.2, the mutational process has two sides. On the one hand, it disturbs balanced molecular mechanisms and thus generates various molecular diseases. On the other hand, it provides new genetic material for selection and acts as a driving force of evolution.

As seen from the above examples of tumors that have already played a role in evolution, a similar "evolution vs. pathology paradox" may also exist for tumors. Tumors could be considered as mutations at the multicellular level. On one hand, tumor processes may lead to various malignancies that are harmful to individual organisms. On the other hand, tumor processes provide excess cell masses with high biosynthetic potential, and this may be used in progressive evolution for the generation of new cell types, tissues and organs, if organisms with tumors survive long enough to leave the progeny.

Chapter 8

The General Principles and Molecular Mechanisms of the Origin of Novel Genes

The hypothesis of evolution by tumor neofunctionalization suggests that expression of evolutionarily novel genes in tumors leads to the appearance of new functions and to the origin of new cell types, tissues and organs in progressive evolution. In this chapter, I will review what is known about the origin and properties of evolutionarily novel genes.

The origin of novel genes is one of the major events in genome evolution. When the origin of evolutionarily novel genes is considered, it is implied that novel genes originate in DNA of germ cells, so that they are inherited. The origin of new genes in somatic or tumor cells does not lead to their inheritance in progeny organisms and is not discussed in this book.

The phenomenon of the origin of novel genes is now well established. Many reviews are devoted to this topic [Eichler, 2001; Long et al., 2003; Zhang, 2003b; Taylor and Raes, 2004; Babushok et al., 2006; Hahn, 2009; Kaessmann et al., 2009; Van de Peer et al., 2009; Innan and Kondrashov, 2010; Kaessmann, 2010]. In this chapter, I will discuss the general principles and molecular mechanisms of the origin of novel genes.

Novel genes may originate from pre-existing genes or *de novo*. It appears that gene duplication, exon shuffling and *de novo* origin are the most fundamental modes, or principles, of the origin of novel genes. Transposons and endogenous retroviruses also participate in gene origin and genome evolution. There are many more molecular mechanisms that participate in the process of gene origin — several molecular mechanisms for each of the major modes of gene origin. It is also generally believed that gene duplication, especially whole genome duplication, is a major determinant of the evolution of complexity and the origin of evolutionary novelties.

Evolution by Tumor Neofunctionalization.

8.1 GENE DUPLICATION

It was the discovery of gene duplication that paved the way for the studies of the genetic basis of progressive evolution. On one side, gene duplication is a major mode of genome evolution. On the other side, "evolution by gene duplication" is a hypothesis that was suggested to explain progressive evolution through gene duplication and divergence. The original hypothesis posited that pre-existing genes are under the control of natural selection, and that their evolution is constrained within their existing function. The extra copy of the existing gene gets out of the control of natural selection, so that accumulation of mutations in this extra copy may lead to the origin of a novel gene with related or even new functions. Gene duplication is considered as providing the "raw material" for the origin of new genes and thereby making possible the evolvability of living organisms (the concept of neofunctionalization, or the classical neofunctionalization model). This concept also suggests that the majority of duplicates become inactive pseudogenes due to degenerative mutations, and only rarely would beneficial mutations lead to the emergence of a new gene with a novel function [Ohno, 1970].

After the discovery in 1911 by Y. Kuwada of chromosome duplication in maize (*Zea mays*), many authors were involved in discovering new evidence and developing the theory of gene duplication. John Haldane and Hermann Muller were among the first to suggest that gene duplication and subsequent divergence could lead to the origin of new genes [Haldane, 1932, 1933; Muller, 1935]. Calvin Bridges discovered one of the first gene duplications in *D. melanogaster* [Bridges, 1936]. The outstanding contributions of R. Goldschmit, A. Gulick, E.B. Lewis, C.W. Metz, A. Serebrovsky, S.G. Stephens and others are reviewed in [Taylor and Raes, 2004]. After the publication of Susumu Ohno's book [Ohno, 1970], studies of gene duplication became the mainstream of molecular evolution research.

Duplicates of single genes arise at a very high rate, on average 0.01 per gene per million years, which is comparable to the nucleotide substitution rate. After a brief period of relaxed selection, the majority of gene duplicates are silenced. The average half-life of a gene duplicate is approximately 4 million years. Michael Lynch and John Conery, the authors of these estimates, conclude that duplicate genes only rarely evolve new functions [Lynch and Conery, 2000]. Evolutionary trajectories of duplicated genes are highly dependent on the effective size of a population. Neofunctionalization is expected to become more important in populations with increasing size [Walsh, 1995; Lynch et al., 2001]. Duplicate genes are lost more slowly in multicellular than in unicellular organisms. The half-life of duplicate genes increases with the genome size [Lynch and

Conery, 2003]. Dying duplicates are more similar to each other than preserved duplicates [Marotta et al., 2012].

Contrary to the high rate of silencing observed for single gene duplicates, the high level of duplicate-gene preservation occurs in polyploid species. This may be explained by preservation of the stoichiometric relationship between gene products supported by stabilizing selection, as discussed in [Lynch and Conery, 2000].

The variation in duplicability is observed among different functional classes of genes. Different functional categories of genes tend to be duplicated either by single-gene duplication or by whole-genome duplication. The functional classes of gene duplicates retained after polyploidization in different species include ribosomal proteins, protein kinases and transcription factors. Some genes can be more easily adapted to novel functions; others may be more duplication-resistant. Genes expressed earlier in development are more duplication-resistant than genes expressed after embryogenesis [Castillo-Davis and Hartl, 2002]. Genome-wide analyses have demonstrated that gene duplications are common in functional classes involved in adaptations, such as transcription factors, regulators of environmental response, transport proteins, cell surface and secreted proteins (reviewed in [Francino, 2005; Conant and Wolfe, 2008]). In yeast, stress-related genes demonstrate many duplications, but growth-related genes are duplication resistant [Wapinski et al., 2007]. Gene duplication happens preferentially in the sparse part of protein interaction networks [Li et al., 2006]. Preferential duplication of conserved genes was described in eukaryotic genomes [Davis and Petrov, 2004]. The study of 1.36 million protein-coding genes from 40 vertebrates, 23 arthropods and 32 fungi has led to the realization that evolutionary modes of "single-copy control" versus "multicopy license" may reflect a major evolutionary regularity [Waterhouse et al., 2010].

In general, the greater proportion of duplicated genes is conserved rather than predicted by the classical neofunctionalization model [Force et al., 1999]. Many new genes and gene families have evolved due to gene duplication throughout the evolutionary history of living organisms. The global studies of completely or nearly completely sequenced genomes of representative bacteria, archaebacteria and eukaryotes suggest that large proportions of genes were generated by gene duplication in all three domains of life [Zhang, 2003b]. In humans, repetitive elements may represent approximately 40% of genomes [Li et al., 2001; Zhang, 2003b]. Extensive genomic duplication took place during early chordate evolution, when many of the vertebrate gene families were formed. Two rounds of whole genome duplication occurred in ancestors of vertebrates and the third in teleost fishes [McLysaght et al., 2002; Van de Peer et al., 2009].

8.1.1 Molecular Mechanisms of Gene Duplication

The DNA-mediated mechanisms of gene duplication include unequal crossing over, tandem, segmental, chromosomal or genome duplications. The resulting gene duplicates may be organized in a tandem, interspersed or polyploid manner. Segmental duplications are large interspersed segments of DNA with high sequence identity (>90%), usually separated by >1Mb of unique sequences [Marques-Bonet et al., 2009a].

There are also RNA-based mechanisms of gene duplication. RNA-based gene duplication, or retroposition, creates duplicate genes by reverse transcription of RNAs from parental genes. RNAs from all categories generate retrosequences that may be exapted as novel genes or regulatory elements [Brosius, 1999]. Retrogenes are most abundant in mammals where long interspersed nuclear elements (LINEs) that provide the enzyme reverse transcriptase for retroposition are widespread. The majority of retrogenes are produced by genes with high levels of germline expression. They often originate from the X chromosome [Betran et al., 2002; Kaessmann et al., 2009]. A new retrogene is intronless, contains a poly(A) tract and may be flanked by short duplicate sequences [Li, 1997; Betran and Long, 2003].

DNA-mediated gene duplication is a more frequent event in genome evolution, while RNA-based gene duplication is more capable of generating genes with novel functions.

The retroposition is less likely to provide expressed daughter retrocopies than segmental DNA duplication because retrocopies do not contain regulatory elements. So, new promoters and enhancers should somehow be recruited for the origin of new genes, and several mechanisms of such recruitment are described. The ways of promoter recruitment by retrogenes is reviewed in [Kaessmann et al., 2009; Kaessmann, 2010]. Retrogenes usually locate on chromosomes different from that of parental genes. For reasons of different location and new promoter recruitment, the transcribed retrogenes are more capable of evolving new expression patterns and novel functional roles than gene duplicates arising from DNA segmental duplication [Kaessmann et al., 2009; Kaessmann, 2010]. The mammalian X chromosome demonstrates extensive retrogene traffic, whereas the autosomes experience lower gene turnover [Emerson et al., 2004].

There are between 8000 and 14,000 processed pseudogenes created by retroposition in the human genome [Torrents et al., 2003; Zhang et al., 2003]. More than a thousand of those have been transcribed. There are also about 120 human functional retrogenes [Vinckenbosch et al., 2006]. At least one functional retrogene per million years originated in the primate lineage that led to humans [Margues et al., 2005]. It is currently supposed that retropseudogenes, like duplicates originated through DNA-mediated mechanisms, might provide the raw material for functionally important evolutionary innovations [Kaessmann et al., 2009]. Several examples of the functional role of retropseudogenes are described [Tam et al., 2008; Watanabe et al., 2008].

8.2 EXON SHUFFLING

Exon shuffling is the other most important mode of the origin of new genes. At least 19% of the exons in the database were involved in exon shuffling, suggesting an important role for exon shuffling in the origin of new genes [Long et al., 1995]. In *Drosophila*, domain rearrangements occur in 39.5% of gene families [Wu et al., 2012].

The principle of exon shuffling is different from that of gene duplication: new genes are created by recombining previously existing protein-coding domains, which leads to the origin of mosaic proteins [Gilbert, 1978; Patthy, 1985, 1996, 2003; Kaessmann et al., 2002; Long et al., 2003]. The first evidence for exon shuffling was obtained from studies on proteases of blood coagulation and fibrinolysis, followed by the discovery of a variety of multidomain proteins in animals (reviewed in [Patthy, 2003]). The correlation between exon-intron organization of the gene and the domain organization of the corresponding protein is most evident in the case of young vertebrate genes, e.g. genes coding for proteases of blood coagulation, fibrinolytic and complement cascades, selectins, cartilage link proteins, fibronectin, factor H, tenascins, etc. [Patthy, 2003].

As far as exon shuffling leads to the generation of new genes, the origin of exons themselves becomes a problem to be addressed. Exon duplications may play a role, but the greater number of new exons, especially in mammals, probably originated through "exonization" of intronic sequences. This process is described in many genomes including human, mouse, dog and fish. New exons frequently appear in these genomes. Comparison of eight vertebrate genomes revealed the birth and evolution of human exons. The majority of new exons turned out to be composed of interspersed highly repeated sequences, especially *Alu* [Zhang and Chasin, 2006].

The fact that most new exons overlap with repeats makes it unlikely that they represent ancient shuffled exons. This was confirmed by comparing new human exons against the mouse EST database and led to the conclusion that most new exons do not arise from exon shuffling. Drawing on these data and considerations, Xiang Zhang and Lawrence Chasin suggested that *de novo* recruitment rather than shuffling is the major route by which new exons are added to genes [Zhang and Chasin, 2006].

8.2.1 Molecular Mechanisms of Exon Shuffling

The mechanisms of exon shuffling include illegitimate recombination [van Rijk et al., 1999, 2003], retroposition [Moran et al., 1999] and segmental duplication [Eichler, 2001]. Exons could be shuffled by L1 retrotransposon-mediated 3′ transduction [Moran et al., 1999] and by atypical splicing, which in an evolutionary perspective may also lead to exon shuffling [Babushok et al., 2006].

Some authors suggest that atypical splicing of existing genes is the most prevalent mechanism of novel protein creation [Babushok et al., 2006]. Atypical splicing would include alternative splicing within the single-gene transcripts and intergenic splicing of transcripts from tandemly located genes. Although splicing events do not alter the number of genes in DNA, transcription-induced chimeras may evolve into gene fusions, and alternative splicing may evolve to gene fission (reviewed in [Babushok et al., 2006]). For instance, the chimeric *PIPSL* gene was formed by L1-mediated retrotransposition of a readthrough, intergenically spliced transcript in hominoids [Babushok et al., 2007]. This phenomenon was called transcription-mediated gene fusion, and many examples (hundreds) of intergenic splicing have been described in the human genome. The authors suggest that it is a novel mechanism of gene origin, where transcription-induced chimerism followed by retroposition may result in a new gene [Akiva et al., 2006]. At least 4–5% of the tandem gene pairs in the human genome can be transcribed into a single RNA coding for chimeric protein [Parra et al., 2006].

Alternative splicing often participates in the exonization process. In the majority of cases the new exon is alternatively spliced and expressed at low levels. In such situations, splice variants with and without the new exon are represented, and the pre-existing function is not destroyed. This opens the way for the origin of new functions and/or new functional modules due to the novel exon, an evolutionary strategy similar to but different from that of gene duplication [Gilbert, 1978; Nekrutenko, 2004; Wang et al., 2005; Sorek, 2007]. The comparison of human, mouse and rat genomes indicates that alternative splicing is associated with an increased frequency of exon creation and/or loss [Modrek and Lee, 2003].

Alu-containing exons are alternatively spliced. Comparative analysis of transposed element insertion within human and mouse genomes reveals *Alu*'s unique role in shaping the human transcriptome. Transposed element exonization may be a source of new constitutively spliced exons [Sorek et al., 2002; Sela et al., 2007].

8.3 *DE NOVO* GENE ORIGIN

At the earliest stages of molecular evolution, even before the origin of the genetic code, when no genes existed yet, various activities of pre-biological polymers were selected among molecules with low activities simultaneously with the selection of primitive organisms. In this way, which may be called "sense acquisition," the first molecular functions and genes might have appeared. This process of junk-to-sense transition may still be operational in the origin of new genes. "Senseless" or "junk" DNA sequences may acquire new functions in the organism and become new genes. In this context, the new functions may be connected not only with protein-coding genes, but also with various functional noncoding RNAs. This mechanism of novel genes origin is called *de novo* origin, or origin "from scratch."

De novo gene origin is considered to be important in the evolution of archaeal and protobacterial gene content [Snel et al., 2002]. Examples of the origin of novel genes from noncoding DNA in eukaryotes are also known and have been increasing during recent years. Several novel genes originating from noncoding DNA, frequently X-linked and exhibiting testis-biased expression, have been initially described in *Drosophila* [Begun et al., 2006; Levine et al., 2006; Begun et al., 2007; Chen et al., 2007]. According to Qi Zhou and co-authors, up to 12% of new genes in *Drosophila* are due to *de novo* origination [Zhou et al., 2008]. After *Drosophila*, *de novo* origin of a new protein-coding gene was described in *Saccharomyces cerevisiae* [Cai et al., 2008], rice [Xiao et al., 2009], and humans [Knowles and McLysaght, 2009]. Three novel human protein-coding genes have been shown to originate from noncoding DNA since the divergence from chimps. These genes have no protein-coding homologs in any other genome. A few human-specific mutations altered protein-coding capacity by destroying "disablers" in the ancestral sequences. The existence of protein-coding genes is supported by expression and proteomic data. The authors estimate that up to 18 novel genes may have originated *de novo* in a human genome [Knowles and McLysaght, 2009]. But a more recent study of *de novo*−originated (since divergence with chimp) human protein-coding genes identified 60 such genes in the human genome, suggesting that *de novo* origin of genes is not extremely rare [Wu et al., 2011]. The lack of orthologous sequences outside of Old World primates suggests *de novo* origin of the *morpheus* gene family, which may be one of the first cases of its kind described in the literature [Johnston et al., 2001; Kaessmann, 2010].

In the house mouse (*Mus musculus*), the emergence of a new gene from an intergenic region has been discovered. The gene has three exons, shows alternative splicing and is expressed in the testis. Its transcript probably functions as noncoding RNA. The emergence of this gene correlates with indel mutations in the 5′ regulatory region, showing that cryptic signals for transcript regulation and processing indeed exist in intergenic regions and can be involved in the *de novo* origin of new genes [Heinen et al., 2009].

It was suggested that *de novo* origin is characteristic of orphan genes (i.e. genes that do not have homologs in other species; see below) [Long, 2007]. Indeed, of 270 primate orphan genes, 15 (5.5%) could have originated *de novo* from noncoding genomic sequences [Toll-Riera et al., 2009]. D. Tautz and T. Domazet-Loso argue that *de novo* origin may be the predominant method for the evolutionary origin of orphan genes [Tautz and Domazet-Loso, 2011].

The studies of *S. cerevisiae* ORFs located in non-genic sequences have detected their widespread translational activity and led to the hypothesis that there is an evolutionary continuum from non-genic sequences to genes that evolve *de novo* through transitory proto-genes [Carvunis et al., 2012].

8.3.1 Molecular Mechanisms of *De Novo* Gene Origin

New promoter elements such as GC-islands, TATA-boxes, LINE1 promoters or retroviral LTRs may arise as a result of mutational process, gene rearrangements, retrotransposition or viral infection. Such events can lead to expression of "senseless" DNA sequences that subsequently may accumulate mutations that alter their protein-coding capacity. The senseless DNA sequences may acquire new functions in the organism and become new genes. Noncoding RNAs may eventually acquire ORF and become protein-coding mRNAs. These could be mechanisms of *de novo* gene origin. Exonization by alternative splicing may be the mechanism of *de novo* exon origin.

8.4 THE ROLE OF TRANSPOSONS IN GENE ORIGIN

Retrotransposons and endogenous retroviruses participate in gene origin and genome evolution. Non-LTR retrotransposons include the long interspersed nuclear element 1 (LINE-1), *Alu* and SVA elements. They collectively make up about one-third of the human genome [Smit, 1996; Lander et al., 2001; Venter et al., 2001] and continue to be active in humans, primates and mammals [reviewed in Cordaux and Batzer, 2009]. SINE repeats, especially *Alu*s, are enriched in segmental duplications breakpoints in primates. The expansion of *Alu* elements in primates made the ancestral human genome susceptible to *Alu-Alu*—mediated rearrangement events which supported the expansion of segmental duplications [Marques-Bonet et al., 2009a]. LINEs may provide the enzyme reverse transcriptase for retroposition in mammals where retrogenes are most abundant [Kaessmann et al., 2009]. LINE1 promoter can also provide regulatory elements for downstream retrogenes [Zaiss and Kloetzel, 1999]. LTRs of endogenous retroviruses may play a similar role [Kowalsky et al., 1997]. Retrotransposon transcription has a key influence upon the transcriptome of mammalian cells. A considerable proportion of the transcripts of the human genome initiates from retrotransposon elements [Faulkner et al., 2009]. Several hundred human genes use transcriptional terminators originated from LTR retroposons [Lander et al., 2001]. L1 and SVA retrotransposons can transduce the downstream genomic sequences [Moran et al., 1999; Xing et al., 2006]. Such 3' transduction has played a role in the emergence of many primate genes [Xing et al., 2006]. Transposons and endogenous retroviruses may directly participate in functional gene creation. The majority of new human exons have turned out to be composed of interspersed highly repeated sequences, especially *Alu* [Zhang and Chasin, 2006]. For example, in his 2004 paper Roy Britten described three functional genes originating from mobile elements [Britten, 2004]. In the case of *AD7C*, which codes for a neuronal thread protein, the coding sequence is 99% made up of a cluster of *Alu* sequences. The coding sequence of gene *BNIP3*, the protein product of which is involved in controlling apoptosis, is 97% made up of sequences from HERV70RM human endogenous retrovirus. *Syncytin*, the third gene listed by Britten, is of special interest because it is a functioning gene in the human placenta. Its coding sequence is made from an endogenous retrovirus HERV-W envelope gene [Lander et al., 2001; Britten, 2004]. *Syncytin* genes originated independently from envelope genes of endogenous retroviruses in primates, rodents and lagomorphs [Mi et al., 2000; Dupressoir et al., 2009; Heidmann et al., 2009]. There are also several examples of *gag* genes domesticated from retrotransposons. Telomerase may have been created through domestication of the reverse transcriptase gene of a non-LTR retrotransposon (reviewed in [Volff, 2006]). As J. Brosius posited, RNAs from all categories generate retrosequences that may be exapted as novel genes or regulatory elements [Brosius, 1999].

DNA-mediated transposons also played a role in the evolution of eukaryotic genomes. They might have done so by generating allelic diversity through insertion, by mediating epigenetic effects on gene expression, by induction of chromosomal rearrangements or by serving as a source of material for the emergence of new genes. DNA transposons were involved in gene transduction, gene duplication and exon shuffling. There are dozens of examples of DNA transposon-derived genes, 43 of which are described in the initial sequencing and analysis of the human genome [Lander et al., 2001]. The origin of V(D)J recombination, a key step in the origin of the adaptive immune system of jawed vertebrates, is attributed to DNA transposon domestication. Finally, transposase DNA-binding domains might participate in the creation of genetic networks (reviewed in [Lander et al., 2001; Volff, 2006; Feschotte and Pritham, 2007]).

8.5 THE ORIGIN OF MULTIGENE FAMILIES

Gene duplication may generate groups of homologous genes with similar structure and functions which often are organized in tandem clusters. These groups of genes are called gene families. The examples are prolactin, histone, tRNA, rRNA, globin and immunoglobulin gene families. Gene families may change in evolution due to differential duplication and loss of genes. There is evidence for adaptive expansion and adaptive contraction of gene families, and for the origin of new families and the death of previously existing families. Multigene families have been implicated in the evolution of complexity; e.g. the expansion and contraction of multigene families played an important role in the independent evolution of mammals and birds. Gene family evolution may have been important for humans because the rate of gene gain and loss is accelerated in primates [Ohta, 1991; Walsh and Stephan, 2001; Fortna et al., 2004; ICGSC, 2004; Vogel and Chothia, 2006; Hahn et al., 2007; Demuth and Hahn, 2009].

Matthew Hahn and co-authors discussed two definitions of gene families, *sensu stricto* and *sensu lato*. The original *sensu stricto* definition was applied to gene paralogs within the genome of a single species. The broader definition of a gene family includes both paralogs within the same species and orthologs and paralogs from other species. The broader definition is connected with the appearance of comparative genomics, with its ability to compare the number of gene copies among species. The broader meaning implies that every gene belongs to a gene family, even single copy genes, and every gene family must first have appeared as a single-copy family [Hahn et al., 2005, 2007b; Demuth and Hahn, 2009].

So, as with single copy genes, the criterion for definition of the novel gene family would be the lack of orthologs in the other species. The presence of the ortholog genes in the closely related species would indicate that the gene family is young.

The origin of new gene families is connected either with rapid divergence from pre-existing families, or with expansion of novel genes originated *de novo* or through other mechanisms. In *Drosophila*, domain rearrangements occur in 35.9% of gene families [Wu et al., 2012]. The expansion of gene number in the families correlates with positive selection at the nucleotide level [Hahn et al., 2007, 2007b]. The frequency of adaptive evolution is higher among duplicates than among orthologs, which supports the role of adaptive evolution in the maintenance of duplicate genes [Han et al., 2009].

For example, substantial expansions of the *morpheus* gene family, supported by positive selection, have occurred in humans and African apes [Johnston et al., 2001]. The lack of orthologous sequences outside of Old World primates suggests *de novo* origin of the *morpheus* gene family [Kaessmann, 2010].

J. Demuth and M. Hahn argue that new families typically originate with orphan genes [Demuth and Hahn, 2009].

8.6 THE ORIGIN OF NONCODING RNA GENES

Up to 70% of the human genome is transcribed, yielding a great amount of so-called "noncoding RNAs" [Birney et al., 2007]. Conventionally, noncoding RNAs (ncRNAs) are divided into small ncRNAs (<200 nucleotides) and long ncRNAs (>200 nucleotides). More than 20 types or individual classes of ncRNAs are described, many of them involved in carcinogenesis (reviewed in [Ponting et al., 2009; Carninci, 2010; Prensner and Chinnaiyan, 2011]).

Noncoding RNA genes may originate from protein-coding genes. The best studied example is the *Xist* lncRNA gene, which evolved in eutherians by pseudogenization of a protein-coding gene and participates in inactivation of a female X chromosome [Duret et al., 2006]. In a similar manner, some primate microRNAs locate within processed pseudogenes [Devor, 2006]. Pseudogene-derived small interfering RNAs regulate gene expression in mammals [Tam et al., 2008; Watanabe et al., 2008].

Sphinx (spx), a young chimeric RNA gene in *Drosophila melanogaster*, has lost its coding ability [Wang et al., 2002]. The long noncoding RNA gene *Pldi* in mice originated *de novo* from an intergenic region [Heinen et al., 2009]. A similar long noncoding multi-exon RNA originated in dogs following intrachromosomal rearrangement. LncRNAs may also originate after the insertion of a transposable element [Ponting et al., 2009].

Long noncoding RNA genes usually do not form large families. Only a few examples of duplicated lncRNA loci are known; these are connected with retrotransposition or local tandem duplications [reviewed in [Ponting et al., 2009]).

On the contrary, clusters of Piwi-interacting RNAs (piRNAs), small non-coding RNAs involved in transposon silencing, originated by duplication based on ectopic (nonallelic homologous) recombination, and their clusters expend in mammals at rates higher than those of any known gene family [Assis and Kondrashov, 2009].

MicroRNAs also expand through duplication events. Studies of miRNAs in human and chimpanzee brains have shown that evolution of miRNAs is ongoing and that there are many emerging miRNAs [Berezikov et al., 2006].

8.7 THE ORIGIN OF NEW GENES IS A WIDESPREAD AND ONGOING PROCESS

Scientists are debating over which modes and mechanisms of gene origin are more predominant. For instance, Babushok and co-authors, in their review of molecular mechanisms of new gene formation, enumerate them in the following order: (1) atypical splicing, both within and between existing genes, selected over time, (2) tandem and interspersed segmental DNA duplications, and (3) retrotransposition events [Babushok et al., 2006]. On the other hand, there is growing awareness that *de novo* gene origin plays a more important role than previously anticipated [Tautz and Domazet-Loso, 2011].

But it looks like different mechanisms take the lead under different circumstances in the evolution of different organisms. In Archaea and Proteobacteria, gene loss, *de novo* origin, gene duplication and horizontal gene transfer (in order of quantitative importance) are the processes that shape the genome during the course of evolution [Snel et al., 2002]. Some organisms use in their evolution certain predominant methods of new gene origin; e.g. tandem gene duplication is used by *Daphnia pulex* [Colbourne et al., 2011]. In *Drosophila*, tandem gene duplication has generated the majority of duplicates limited to *D. melanogaster* or *D. yakuba*; dispersed duplicates are most abundant among new genes that are functional and shared by multiple species; *de novo*—formed genes comprise 12% of all new genes; retroposition has generated 10% of new genes; and 30% of new genes have formed chimeric gene structures [Zhou et al., 2008].

Finally, the combination of several mechanisms may take place in the origin of novel genes, as in the case of the *jingwei* gene in *Drosophila*, the origin of which combined exon shuffling, segmental duplication and retroposition [Long et al., 2003]; *sphinx*, a young chimeric RNA gene in *Drosophila*, originated as an insertion of a retroposed sequence which recruited a nearby exon and intron [Wang et al., 2002]; and the *PIPSL* chimeric gene in hominoids, the origin of which combined readthrough transcription, intergenic splicing and L1 retrotransposition [Babushok et al., 2007]. There are many other examples. The antifreeze glucoprotein gene of Antarctic notothenioid fish originated by recruitment of the 5′ and 3′ ends of an ancestral trypsinogen gene and *de novo* amplification of a 9-nucleotide Thr-Ala-Ala coding sequence [Chen et al., 1997]. A novel primate gene family *RGP* descended from the highly conserved nucleoprotein RanBP2 gene by many genetic rearrangements including segmental duplications, inversions, translocations, exon loss and domain accretion [Ciccarelli et al., 2005].

The general conclusion that authors make after considering different principles and mechanisms of the origin of novel genes is that this process is frequent, ongoing and widespread [Lynch et al., 2001; Zhang, 2003b; Babushok et al., 2006]. The emerging picture is that genomes are plastic, evolving entities; similes like "structural fluidity" [Eichler, 2001], "genomes in flux" [Snel et al., 2002], "genomes... as proving grounds" [Wolfe and Li, 2003] or "the life and death of gene families" [Demuth and Hahn, 2009] are used to describe the picture of evolving genomes.

Chapter 9

The Origin of Evolutionarily Novel Genes and Evolution of New Functions and Structural Complexity in Multicellular Organisms

In this chapter the relation of evolutionarily novel genes to functional and morphological evolution of organisms will be discussed. I will examine the concept (which is now receiving more and more supporting evidence) that progressive evolution (i.e. the origin of morphological novelties and evolutionary innovations and the increase in complexity of multicellular organisms) is connected with the origin of evolutionarily novel genes, i.e. genes with altered or new functions as compared to pre-existing genes. This concept was elaborated by the collective effort of many scientists in the twentieth century.

9.1 NEW AND ALTERED FUNCTIONS OF NOVEL GENES

Different modes of gene origin — gene duplication, exon shuffling and *de novo* origin — are connected with somewhat different ways of acquisition of new functions.

The evolutionary fate of duplicate genes can be pseudogenization, conservation and increase of gene function (e.g. ribosomal proteins, rRNAs and histone genes); neofunctionalization, i.e. the origin of an entirely new function; and subfunctionalization, when each of two genes retains only a subset of the original functions [Ohno, 1970; Force et al., 1999; Zhang, 2003b; Hahn, 2009].

Pseudogenes can make up a considerable portion of the genome in multicellular organisms [Demuth and Hahn, 2009]. For example, there are between 8000 and 14,000 processed pseudogenes created by retroposition in the human genome and several thousand that arose by segmental duplication (nonprocessed pseudogenes) [Torrents et al., 2003; Zhang et al., 2003]. The human genome may contain more pseudogenes than normal genes [Torrents

Evolution by Tumor Neofunctionalization.

et al., 2003]. Up to one-third of all pseudogenes are transcribed. At least some pseudogenes may be functional [Betran et al., 2002b; Krasnov et al., 2005]. Transcribed noncoding pseudogene RNAs may have a function in regulating the normal genes from which they originate [Babushok et al., 2007; Tam et al., 2008; Watanabe et al., 2008].

Neofunctionalization is the acquisition of an entirely new function by a gene duplicate, although in many cases the gene duplicate evolves not a completely new, but related function. Theoretical studies show that, for sufficiently large populations, a new gene function driven by selective advantage is the expected fate of most duplicated genes [Walsh, 1995; Lynch et al., 2001].

The number of examples of neofunctionalization is increasing. These examples include convergent neofunctionalization by positive Darwinian selection after ancient recurrent duplications of the xanthine dehydrogenase gene [Rodriguez-Trellers et al., 2003]; neurally expressed locus of triosephosphate isomerase directionally selected following gene duplication [Merrit and Quatro, 2001]; and *Dntf-2r*, a young positively selected *Drosophila* retroposed gene with specific male expression [Betran and Long, 2003]. Patterns of rate asymmetry and gene loss strongly suggest widespread neofunctionalization of yeast genes soon after whole-genome duplication [Byrne and Wolfe, 2007].

The evolution of an antifreeze glycoprotein gene from a trypsinogen gene was described in Antarctic notothenioid fish. The antifreeze gene originated by recruitment of the 5' and 3' ends of an ancestral trypsinogen gene and *de novo* amplification of a 9-nucleotide Thr-Ala-Ala coding sequence. This is an example of the origin of a new gene with entirely new functions and adaptations which occurred 7–15 million years ago coincidentally with the freezing of the Antarctic Ocean [Chen et al., 1997]. In an Antarctic zoarcid fish, evolution of an antifreeze protein from an old sialic acid synthase gene by neofunctionalization under escape from adaptive conflict was described [Deng et al., 2010].

An interesting case of evolution of a new function is the *RNase1B* duplicate of the pancreatic ribonuclease gene (*RNase1*) in Asian colobine monkey, douc langur (*Pygathrix nemaeus*). The *RNase1B* gene evolved more rapidly, than *RNase1* and obtained a different catalytic optimal pH and higher efficiency of degrading RNA. This provided colobine monkeys with the capacity to use leaves as a primary food source and thus adapt to a new nutritional niche [Zhang et al., 2002a]. Similar evolutionary changes have occurred independently in the African leaf-eating monkey's RNase1 gene [Zhang, 2006].

The gene for eosinophil cationic protein (ECP) in primates belongs to the ribonuclease gene family. It originated by gene duplication and positive Darwinian selection operating in the early stage of its evolution [Zhang et al., 1998].

Human-specific duplicates evolving under adaptive natural selection determine novel neuronal and cognitive functions [Han et al., 2009].

There are neofunctionalizations common to all chordates (aldehyde oxidases), vertebrates (retinoic acid receptors, *Myb* genes) and primates (opsins, chorionic gonadotropin, glutamate dehydrogenase) (see [Hahn, 2009] for references).

The enzymatic mechanism used to build the innovative function is usually conserved, as in the cases of the glutamate dehydrogenase gene in humans and apes (*GLUD2*), the *jingwei* gene in *Drosophila* and the gonadal paralog of the pig cytochrome *P450arom* gene (reviewed in [Conant and Wolfe, 2008]).

According to an adaptive radiation model, the evolution of a new function may start with adaptation by amplification of a gene with primordial function, followed by gene competition, preservation of the most effective gene copy and the pseudogenization and loss of all others [Francino, 2005; Bergthorsson et al., 2007]. The bursts of gene amplification, early positive selection and generation of numerous pseudogenes are considered as evidence supporting the adaptive radiation model. The evolution of the olfactory receptor gene family is considered as a case of adaptive radiation [Francino, 2005].

The subfunctionalization model suggests that each of the duplicate genes performs only a subset of the functions of the ancestral single copy gene and loses a different subcomponent of its function [Force et al., 1999; Prince and Picket, 2002; Conant and Wolfe, 2008]. The possibility of subfunctionalization stems from the multifunctional nature of genes and proteins. Most proteins have a range of activities, often called promiscuous or moonlighting activities [Jensen, 1976; Copley, 2003; Jeffery, 2009]. Promiscuous protein functions can evolve under selective conditions [Aharoni et al., 2005] or due to neutral genetic drift [Bloom et al., 2007] without losing their original functions. After acquisition of a new function, gene duplication and subfunctionalization may follow. The first described example of this kind (called "gene sharing" by the authors) is lens crystalline genes in which the recruitment of new functions and enhanced expression in the lens preceded gene duplication [Piatigorsky and Wistow, 1991]. Gene duplications facilitated the evolution of beta-catenins with novel functions and allowed the evolution of multiple, single-function proteins (cell adhesion or signaling) from the ancestral, dual-function protein [Schneider et al., 2003]. *Fgf8/17/18*, a single gene of hemichordate *Saccoglossus kowalevskii*, is homologous to vertebrate *Fgf8*, *Fgf17* and *Fgf18* genes [Pani et al., 2012].

The other type of subfunctionalization is regulatory subfunctionalization, which leads to differential gene expression after duplication. It is described by the duplication-degeneration-complementation (DDC) model [Force et al., 1999]. The authors discussed divergence in expression of duplicate *engrailed* genes in zebrafish. The *eng1* gene of zebrafish is expressed in the pectoral appendage bud, and *eng1b* gene is expressed in the hindbrain/spinal cord, while the orthologous *En1* gene of the mouse is expressed in both places [Force et al., 1999]. So, the ancestor fish gene was probably expressed in two tissues, and each of two duplicates is expressed in only one tissue due to changes in *cis*-regulatory elements. As the authors point out, the DDC model shows how degenerative mutations in regulatory subfunctions support the retention of duplicate genes while the classical model suggests that degenerative mutations lead only to gene loss [Force et al., 1999]. Subfunctionalization generates genes with altered rather than new functions, but it may facilitate neofunctionalization by extending the period of duplicate survival and thus increasing the probability of neofunctionalizing mutation [Conant and Wolfe, 2008]. A new term, "subneofunctionalization," was coined for designation of situations in which subfunctionalization was followed by substantial neofunctionalization [He and Zhang, 2005], and subfunctionalization is considered as a transition state to neofunctionalization [Rastogi and Liberles, 2005].

New subcellular localization of duplicate genes' protein products may be an important factor of their functional diversification. In *Saccharomyces cerevisiae*, 24–37% of the duplicate gene pairs originated from a whole-genome duplication encode proteins that localize to distinct cellular compartments [Margues et al., 2008]. Hominid-specific *CDC14Bretro* encodes four different splice isoforms with distinct subcellular localizations and functional properties [Rosso et al., 2008a]. Hominoid-specific *GLUD2* genes code for the glutamate dehydrogenase enzyme specifically targeted to mitochondria [Rosso et al., 2008b]. Henrik Kaessmann and co-authors suggest that rapid, selectively driven subcellular adaptation may be a general principle of the emergence of new gene functions [Margues et al., 2008; Rosso et al., 2008a,b]. This idea was also introduced by other authors [Byun-McKay and Geeta, 2007].

Exon shuffling evidently would lead to novel combinations of already-existing functions. It played an important role in the evolution of the modular proteins involved in cell communications in Metazoa. Novel functions in Metazoa include those related to cell–cell interaction and adhesion. Proteins that participate in the ancient mode of cell adhesion in animals, attachment to basement membranes, are common to hydra, worms, flies and vertebrates (laminins, type IV collagen and others). Many vertebrate transmembrane proteins that take part in cell adhesion are also found in the worm and fly (cadherins and many others). The majority of signaling pathways are present in flies, worms and vertebrates. But there are also genes and proteins that participate in cell adhesion and cellular interactions that emerged in deuterostome/chordate lineage: the modular collagens typical of vertebrates; novel types of integrins; the cartilage aggregating proteoglycan and other members of this family; thrombospondins and the related cartilage oligomeric matrix protein; the majority of blood proteins associated with hemostasis and defense mechanisms; novel types of cadherins and other proteins expressed in the brain; etc. These are novel multidomain proteins that originated in vertebrates by exon shuffling and modular assembly [Patthy, 2003].

It is assumed that "chimeric" or fusion genes evolve new functions because of considerable structural differences with their parental genes, and indeed there are many such examples (reviewed in [Hahn, 2009]). Many novel chimeric genes evolve under positive selection which suggests the involvement of adaptive models of neofunctionalization [Hahn, 2009]. In the evolution of chimeric fusion genes, rapid shifting away from ancestral functions (neofunctionalization) may be a general phenomenon, as shown by the study of *jingwei*, *Adh-Finnegan* and *Adh-Twain* genes in *Drosophila* [Jones and Begun, 2005].

De novo—originated genes should have some novel, previously nonexistent functions. These functions supposedly should fit into developmental or terminal differentiation classes of functions. In rice, a *de novo*—originated gene participates in the pathogen-induced defense response [Xiao et al., 2009]. In *Drosophila*, novel genes derived from noncoding DNA often exhibit testis-biased expression and probably have functions related to male reproduction [Levine et al., 2006]. The novel gene that emerged from an intergenic region in mice may have a similar function [Heinen et al., 2009]. The analysis of human *de novo*—originated genes has shown their possible functional role in the brain and testes [Li et al., 2010; Wu et al., 2011]. The low level of expression of human *de novo*—originated young genes suggests that their function is not yet well established [Wu et al., 2011]. Although the number of examples is still small, the impression is that *de novo*—originated genes may be connected with new or modified differentiation functions. This conclusion is in agreement with the different types of approaches that show that latter ontogenetic stages are characterized by the expression of newly evolved genes [Domazet-Loso and Tautz, 2010b].

Complex genomic rearrangements may be needed to generate novel gene functions. A novel primate gene family *RGP* descended from the highly conserved nucleoprotein *RanBP2* gene by many genetic rearrangements including segmental duplications, inversions, translocations, exon loss, and domain accretion [Ciccarelli et al., 2005].

The analysis of gene family evolution in yeast, flies and mammals has discovered common functional categories of rapidly evolving families, i.e. immune defense/stress response, metabolism, cell signaling, chemoreception and reproduction-related families. Gene families related to infection and disease demonstrate a consistent pattern of expansion. Genes with these functions also include the most rapidly evolving genes in *Drosophila* and mammals. In mammals, there is also expansion of the genes involved in olfaction (reviewed in [Babushok et al., 2006; Demuth and Hahn, 2009]). The analysis of segmental duplications in humans has shown particular enrichment of the genes associated with immunity and defense, membrane surface interactions, drug detoxification and growth/development [Bailey et al., 2002]. A gene of *de novo* origin negatively regulates the pathogen-induced defense response in rice [Xiao et al., 2009].

Many new genes, originated by different modes and mechanisms, acquire functions necessary for spermatogenesis, which is in agreement with their expression in testis [McCarrey and Thomas, 1987; Kleene et al., 1998; Betran and Long, 2003; Paulding et al., 2003; Zhang et al., 2003; Bradley et al., 2004; Margues et al., 2005; Vinckenbosch et al., 2006; Babushok et al., 2007; Kaessmann et al., 2009; Kaessmann, 2010; Wu et al., 2011].

The variety of adaptive functions due to novel genes includes even the evolution of courtship behavior through the origination of the *sphinx* gene in *Drosophila* [Dai et al., 2008].

New genes quickly become essential, i.e. they rapidly evolve essential functions and participate in development, as shown by studies in *Drosophila*. In these studies, the expression of 195 young genes was knocked down with RNA interference, showing that 30% of them are essential for viability [Chen et al., 2010]. A majority of the functional new genes in *Drosophila* are dispersed duplicates [Zhou et al., 2008].

Novel combinations of interacting genes may participate in the origin of new functions, which may arise in multidimensional networks of functional interactions, such as the protein–protein interactions [Wagner, 2001; Hughes, 2005; Colbourne et al., 2011].

The conclusion of this part may be that the origin and evolution of new genes is connected with the origin and evolution of new molecular and organismal functions. As far as new functions are beneficial to organisms, natural selection will support and enhance such new functions.

9.1.1 Adaptive Evolution and Positive Selection of Novel Genes

Several different models for the evolution and maintenance of gene duplicates have been developed by different authors. In the paper of H. Innan and F. Kondrashov they are classified into several categories which differ in selective forces and evolutionary events at different stages of the duplication history [Innan and Kondrashov, 2010]. The first category includes models that suppose that fixation of the duplicate is a neutral process. The subfunctionalization model falls within this category. According to the subfunctionalization model, the longer retention of duplicate genes and the increase in gene number in multicellular organisms may be a result of neutral evolution during the process of subfunctionalization, and may not be driven by adaptive processes [Lynch and Conery, 2003].

The models of the second category consider situations in which duplications are advantageous (beneficial increases in dosage, shielding against deleterious mutations or the immediate emergence of new functions). In these models, positive selection supports duplication. The third category of models deals with duplication in genes with genetic polymorphisms. In these models, positive selection on pre-duplication variation takes place. The last category includes the dosage balance model based on whole genome duplication where the fixation phase is absent [Innan and Kondrashov, 2010].

Innan and Kondrashov put the classical neofunctionalization model of Ohno [Ohno, 1970] in the first category, suggesting that fixation of the duplicated copy is a neutral process, but positive selection can target the derived copy during the preservation stage [Innan and Kondrashov, 2010].

M. Hahn specifically discussed two models for neofunctionalization which differ in the role of adaptive natural selection. The neutral model proposes that mutations at the redundant locus are not fixed by selection but accumulate due to drift and later may be used by natural selection under appropriate conditions. The adaptation model posits that neofunctionalization occurs by adaptive fixation of mutations, although it does not exactly specify how early in the process of new gene origin selection starts. Hahn points out the difficulty in distinguishing between neutral and adaptation models for neofunctionalization, and in distinguishing between neofunctionalization models and subfunctionalization models with patterns of positive selection, e.g. gene sharing. The same is true for other major models of maintenance of gene duplicates which offer overlapping predictions. Both functional and evolutionary data are necessary to distinguish between models and outcomes, i.e. neofunctionalization and subfunctionalization [Hahn, 2009].

Several models mentioned above suggest that positive Darwinian selection may participate in the evolution and maintenance of the novel genes. The increased rate of gene evolution due to positive selection is difficult to distinguish from the relaxation of selective constraints after duplication. There have been criticisms and discussion of the positive selection/adaptive nature of gene evolution, e.g. of *ASPM* and *MCPH1* genes [Evans et al., 2005; Mekel-Bobrov et al., 2005; Currat et al., 2006; Mekel-Bobrov et al., 2006; Mekel-Bobrov and Lahn, 2007; Timpson et al., 2007; Yu et al., 2007 Hahn, 2008]. Nevertheless, comparing the rate of amino acid replacement substitution with the rate of synonymous substitution, a population genetic analysis of polymorphisms and the findings of convergent evolution support the adaptive evolution of the novel genes [McDonald and Kreitman, 1991; Kreitman, 1996; Long et al., 2003]. The criterion $K_A/K_S > 1$ is one of the most stringent tests of selection, and is characteristic of a significant proportion of young duplicates [Hahn, 2009]. Positive Darwinian selection is supposed to be a driving force in the evolution of the novel genes in many cases, supporting the appearance of new adaptive functions [Hughes, 1994; Moore and Purugganan, 2003; Innan and Kondrashov, 2010]. Two types of positive selection pressures are described: directional selection, which leads to neofunctionalization, and diversifying selection, which does not lead to change of the gene function [Zhang and Rosenberg, 2002].

There are many examples of rapidly evolving novel genes supported by positive selection (reviewed in [Long et al., 2003; Wolfe and Li, 2003]): *jingwei* in *Drosophila* [Long and Langley, 1993], lysozyme and ribonuclease genes in primates [Messier and Stewart, 1997; Zhang et al., 1998, 2002a], human *FOXP2* [Zhang et al., 2002b]; etc. (see the discussion of neofunctionalization above).

Newer exons evolve at a higher rate compared with older exons, as evidenced by the higher K_s, K_a and K_a/K_s ratio, as well as the SNP density [Zhang and Chasin, 2006]. Different protein domains (and their corresponding exons) can evolve with different evolutionary rates. Both conserved and rapidly evolving regions have been described in the BRCA1 protein [Pavlicek et al., 2004], ASPM protein [Kouprina et al., 2004a] and many other proteins, reflecting a mosaic of positive and negative selection. Rapid subcellular adaptation of gene duplicates protein products driven by positive selection may be a general principle of the emergence of new gene functions [Byun-McKay and Geeta, 2007; Margues et al., 2008; Rosso et al., 2008a,b]. There is also an inverse relationship between evolutionary rate and age of mammalian genes, which may be explained by lower functional constraints at the time of origin of novel genes [Alba and Castresana, 2005, 2007].

In humans, strong positive selection and accelerated evolution was documented for the lactase gene, consistent with the shift to agrarian society and dairy farming [Bersaglieri et al., 2004] and for many other genes with different molecular functions, e.g. transcription factors, genes involved in nuclear transport, sensory perception, immune defenses, tumor suppression and apoptosis [Clark et al., 2003; Bustamante et al., 2005; Nielsen et al., 2005].

More examples of positive selection of genes and proteins can be presented (see e.g. Table 1 in [Wolfe and Li, 2003], Tables 1 and 2 in [Clark et al., 2003] and Table 1 in [Bustamante et al., 2005]).

The frequency of positive selection is much higher among duplicated genes than among orthologs [Han et al., 2009]. Earlier, Tomoko Ohta, the author of the nearly neutral theory of molecular evolution, analyzed a population genetics model for the origin of gene families and suggested that positive Darwinian selection is necessary for the evolution of gene families with novel functions and the evolution of complexity [Ohta, 1991, 1992]. Examples of positively selected gene families are numerous, including orphan genes in *Drosophila* [Domazet-Loso and Tautz, 2003], *morpheus* [Johnson et al., 2001] and *SPANX* gene families in African great apes and hominids [Kouprina et al., 2004b]. The highest rate of piRNA cluster expansion in mammals, as compared to other known gene families, is explained by positive selection which is caused by the need to silence the expanding mammalian transposon families [Assis and Kondrashov, 2009].

Gene gain and loss is accelerated in primates relative to other mammals. Several gene families have expanded or contracted rapidly in primates, including brain-related families in humans. Many such families show evidence of positive selection [Hahn et al., 2007]. Hominid- and human-specific gene families under positive selection are presented in Tables 2 and 3 in [Han et al., 2009].

Human genome-wide studies suggest that ongoing positive selection has been strongest for immune response, reproduction and sensory perception genes. A disproportionately large number of X-linked genes are represented among rapidly evolving genes. Strong signals of recent positive selection are associated with various types of genes related to morphology. Other selected alleles are connected with host−pathogen interactions, DNA metabolism/cell cycle, protein metabolism and neuronal functions, pigmentation pathways, dystrophin protein complex, heat shock proteins, etc. [Sabeti et al., 2006, 2007; Voight et al., 2006; Wang et al., 2006; Williamson et al., 2007; Pickrell et al., 2009]. Many promoter regions of human neural- and nutrition-related genes have undergone positive selection [Haygood et al., 2007]. However, concordance between studies is variable [Hahn, 2007; Nielsen et al., 2007]. Interestingly, more structural genes underwent positive selection in chimp evolution than in human evolution. Human and chimp positively selected genes have different functions [Bakewell et al., 2007]. The proportion of positively selected genes is significantly higher in younger genes in humans, i.e. positive selection may play a role in the faster evolution of younger genes [Cai and Petrov, 2010].

Many examples of rapid evolution and positive selection of new genes described in the literature points out that this phenomenon is widespread. It supports the involvement of novel genes and gene families in adaptation and speciation and in the evolution and enhancement of new functions. The evidence discussed above and other data determined the movement towards a selection theory of molecular evolution that will include a much greater role for natural selection in gene evolution [Hahn, 2008].

9.1.2 Orphan or Taxonomically Restricted Genes

Orphan genes are those that have no homologs in other species. The operational definition, i.e. "coding regions without matches to other genes in the database," depends on the statistics, the size of the database and the species' representation in it [Domazet-Loso and Tautz, 2003]. Lineage-specific gene families may also be called orphans [Hahn et al., 2007b].

Different genomes studied so far contain large numbers (10−20%) of orphan genes. Supposedly orphan genes may play species-specific adaptive roles.

De novo evolution out of noncoding DNA may be the predominant way of origin of orphan genes [Tautz and Domazet-Loso, 2011].

An evolutionary analysis of *Drosophila* proteome has shown that between 26% and 29% of proteins do not have homology with noninsect sequences. A cDNA library from adults contains two times more orphan genes than the cDNA library from *Drosophila* embryos. The authors consider three possible mechanisms of the origin of *Drosophila* orphans: *de novo* origin, the loss of an ancestrally shared gene in other species and the quick evolution of a gene that causes the lack of similarity with ancestral genes. Indeed, they were able to show that most orphans in *Drosophila* evolve more than three times faster than nonorphan genes, although some of them have substitution rates comparable to highly conserved genes. Slow-evolving orphan genes may represent species-specific adaptations [Domazet-Loso and Tautz, 2003]. Tandem gene duplication generated approximately 80% of duplicates that are found in single *Drosophila* species [Zhou et al., 2008].

Similarly to *Drosophila*, in *Daphnia pulex* more than a third of genes have no detectable homologs in any other available proteome. *D. pulex* has more than 30,000 genes that originated due to a high rate of gene duplication and are organized in tandem gene clusters. The most amplified gene families are specific to the *Daphnia* lineage [Colbourne et al., 2011].

A systematic study of primate orphan genes has shown their higher tissue specificity, more rapid evolution and a shorter peptide length. Around 24% of them are highly divergent members of mammalian protein families, 53% contain sequences derived from transposable elements and are located in primate-specific genomic regions and 5.5% might have originated *de novo* from mammalian noncoding genomic regions [Toll-Riera et al., 2009].

The evidence from coelenterate *Hydra* suggests that its orphans participate in the origin of phylum-specific novelties, such as cnidocytes, in the generation of morphological diversity and in the innate defense system [Khalturin et al., 2009].

9.1.3 The Role of Expression in the Origin and Evolution of New Genes and their Functions

Evolution at the level of DNA only is not sufficient for the emergence of the novel gene with a new function. It must be accompanied by expression of RNA and/or protein products and selection for their function in the organism. We have seen that such a situation exists in the origin of new genes through retroposition (see above). The high rate and promiscuous character of expression in the testis is supposed to be important for the role of the testis as a catalyst for the birth and evolution of new genes [Kaessmann, 2010]. There are other examples that illustrate the importance of the expression stage for the origin of new genes.

RNAs from all categories generate retrosequences that may evolve to novel genes [Brosius, 1999]. Ribosomal proteins have the largest number of processed pseudogenes in the human genome, which is connected to their high mRNA expression level [Zhang et al., 2003].

The recruitment of enzymes as crystallins illustrates a situation in which changes in expression occur before gene duplication [Piatigorsky and Wistow, 1991]. Subfunctionalization often leads to the expression of duplicate genes in different cell types [Force et al., 1999].

In yeast *Saccharomyces cerevisiae*, a new *de novo* protein-coding gene *BSC4* was identified. It may participate in the DNA repair pathway.

Corresponding noncoding sequences in *S. paradoxus*, *S. mikatae*, and *S. bayanus* also transcribe, which suggests the importance of transcription stage for the origin of new genes from noncoding sequences [Cai et al., 2008].

In mice, the emergence of a new lncRNA gene from an intergenic region has been described [Heinen et al., 2009]. The gene has three exons, alternative splicing and a specific expression and function in the testis. The region is present in mammals and humans, and experienced a recent selective sweep in *M.m. musculus* populations.

In *Xenopus*, two duplicate genes that participate in D-V patterning — the ventral gene *BMP4* and its dorsal counterpart *ADMP* — have been placed under opposing transcriptional control that determines the D-V pattern formation [De Robertis, 2006]. In *Daphnia pulex*, with its over 30,000 genes organized in tandem clusters, the paralogs often acquire different expression patterns at, or soon after, the time of duplication. Evolution of *hydra*, a recently *de novo*—evolved testis-expressed gene of *Drosophila*, shows a dramatic change in expression level (>20-fold) between *D. melanogaster* and *D. simulans*. Recurrent evolution of both gene structure and expression level may thus be characteristics of newly evolved genes, the authors conclude [Chen et al., 2007].

Highly expressed genes in yeast evolve slowly [Pal et al., 2001]. The same is true for vertebrates [Subramanian and Kumar, 2004]. In the mouse, the rates of evolution of *Pax2*, *Pax5* and *Pax8* inversely correlate with the duration of genes' expression and the importance of their function [Aburomia et al., 2003]. In *Drosophila*, the nematode *Caenorhabditis elegans* and zebrafish, the highest rates of gene evolution (associated with positive selection) are characteristic to genes expressed in adults after they reach reproductive maturity, especially in male gonads [Domazet-Loso and Tautz, 2003; Cutter and Ward, 2005; Davis et al., 2005; Artieri et al., 2009; Garfield and Wray, 2009; Domazet-Loso and Tautz, 2010b]. In mammals, substitution rates at nonsynonymous sites are higher in tissue-specific than in ubiquitous genes [Duret and Mouchiroud, 2000].

Divergence in expression between duplicate genes seems to be a general phenomenon and happens quickly after gene duplication [Wagner, 2000; Gu et al., 2002]. The expression divergence, regulatory- and coding-sequence divergence increase with the age of duplicate genes. By expression divergence, gene duplication makes possible the evolution of tissue and cell type specialization [Li et al., 2005]. A greater proportion of Precambrian genes are broadly expressed (>10 tissues) as compared to the vertebrate-specific genes [Subramanian and Kumar, 2004].

In *Drosophila melanogaster*, the gain of novel expression patterns is rare in comparison with changes in transcript abundance or losses of patterns [Rebeiz et al., 2011].

The expression levels of *de novo*—originated genes in humans are very low, and they are mainly expressed in the cerebral cortex and testes

[Wu et al., 2011]. Rapid relative change in gene expression in the human brain has occurred in parallel with a decreased rate of coding sequence evolution. Genes with expression restricted to the brain show greater conservation than genes expressed in the brain along with other tissues. The evidence suggests that gene-expression differences have played an important role in primate evolution and adaptation [Wang et al., 2007; Marques-Bonet et al., 2009b].

Studies of the evolution of gene expression in mammalian organs using an RNA sequencing (RNA-seq) approach have shown that transcriptome change was slow in nervous tissues and rapid in testes, and slower in rodents than in apes. Closely related species have more similar expression levels [Brawand et al., 2011].

There could be energy constraints on the evolution of gene expression after duplication. A doubling of gene expression after a duplication event is selected against, as shown in yeast *Saccharomyces cerevisiae* [Wagner, 2005]. Earlier I argued that gene competition for space and resources might negatively affect the expression of novel genes [Kozlov 1979, 1996]. That is why excessive tumor cells, functionally unnecessary to the organism, might be instrumental in facilitating the expression of evolutionarily novel genes and their novel functions (see discussion of this issue in Part 10.8).

9.2 NOVEL GENES AND THE EMERGENCE OF EVOLUTIONARY INNOVATIONS AND MORPHOLOGICAL NOVELTIES IN MULTICELLULAR ORGANISMS

Cell differentiation has been a primary feature of the multicellular organisms since their origin. Differentiation gene batteries are the oldest components of the genomes of multicellular organisms [Erwin and Davidson, 2002]. The number of cell types in multicellular organisms has been increasing throughout evolution. The same is true for the number of structural gene families and the increase in gene copy number in such families: gene families expand in size and number [Demuth and Hahn, 2009]. The number of differentiated cell types in the multicellular organism is one of the major characteristics of its complexity [Kozlov, 1979]. There are multicellular organisms with 2, 3, 4, 5 etc. cell types. Placozoan *Trichoplax adhaerens* has 4 cell types, sponges have 16 cell types, cnidarians from 3 to 22 cell types. Chordates demonstrate a burst of the number of cell types which exceeds 100. The detailed account of the number of cell types in different phyla of multicellular organisms is presented in [Bell and Mooers, 1997]. Humans have 411 different specialized cell types including 145 types of neurons [Vickaryous and Hall, 2006], and regulatory systems and structural gene families necessary to determine these differentiated cell types. According to other estimations, only the neuronal population may account for 10,000 different subtypes of neurons in humans [Muotri and Gage, 2006]. There are gene superfamilies whose sizes are strongly correlated with the number of cell types in the organisms [Vogel and Clothia, 2006].

Up to 479 morphological characters can be scored that may be an index of vertebrate morphological complexity. The original vertebrate innovations were first acquired on the chordate stem lineage (as compared to amphioxus and tunicates) and include the neural crest and its derivatives, elaborated placode-derived structures, elaborated brains with rhombometric segmentation, cartilage and possibly mineralization and the axial and the head skeleton. Other innovative characters such as paired appendages, hinged jaws, an adaptive immune system and specialized axial skeleton were acquired later along the jawed vertebrate stem lineage [Aburomia et al., 2003].

The diversity of domain combinations increases with the organism's complexity: the multidomain proteins and signaling pathways of the fly and worm are far more complex than those of yeast [Rubin et al., 2000]. Human genes are even more complex, and more alternative splicing events generate a larger number of protein products. The human proteome is more complex than that of invertebrates due to additional vertebrate-specific protein domains, the richer collection of domain architectures and greater number of multidomain proteins [Lander et al., 2001; Li et al., 2001; Venter et al., 2001]. Multidomain proteins may be connected with an increased organism complexity due to their ability to participate in multiple interactions with many partners, i.e. higher connectivity [Patthy, 2003]. There is strong evidence that exon shuffling played a major role in the origin of the modular proteins involved in extracellular communications of metazoa and continues to be a major source of evolutionary novelty during vertebrate evolution [Patthy, 2003]. Exonization of *Alus* may have promoted the increase of complexity in humans [Sorek et al., 2004; Sela et al., 2007]. The same may be true for other classes of repeats and to other mammals [Zhang and Chasin, 2006]. Insertion of transposed elements into the first and last exons of a gene is more frequent in humans as compared to mice and results in longer human exons [Sela et al., 2007].

Many authors associate evolutionary novelties and an increase of morphological complexity in vertebrates with an increase in genome size and gene number due to gene duplication [Shimeld and Holland, 2000; Aburomia et al., 2003; Freeling and Thomas, 2006; Khaner, 2007; Wagner, 2008]. Gene families expanded by gene duplication in vertebrates encode transcription factors (Hox, ParaHox, En, Otx, Msx, Pax, Dlx, HNF3, bHLH), signaling molecules (hh, IGF, BMP) and structural proteins and enzymes (dystrophin, cholinesterase, actin and keratin). An increase in the number of genes by itself is considered a vertebrate innovation [Shimeld and Holland, 2000].

Of genes expressed in the neural crest, 9% are vertebrate innovations. These new genes originated either by gene divergence after duplication or by *de novo* gene evolution. They are connected with the appearance of novel molecular and cellular functions, and with tissues considered to be vertebrate novelties [Martinez-Morales et al., 2007].

Gene duplication in the *Pax2/5/8* gene family is involved in patterning of the brain and the formation of cerebellum [Aburomia et al., 2003; Khaner, 2007].

A network of transcription factors is involved in the evolution and development of the heart. This network includes *NK2, MEF2, GATA, Tbx* and *Hand* genes. In the vertebrate lineage, from cephalochordates to amniotes through fish and amphibians, the number of heart chambers has increased from one to four. This was paralleled by an increase in the number of genes in the transcription network — from five in cephalochordates to more than 18 in amniotes — through gene duplication. There are two copies of *Hand* gene in amniotes while in cephalochordates, fish and amphibians there is only one. In mice, *Hand1* controls left ventricle formation, and *Hand2* controls right ventricle formation. Therefore there is correlation between *Hand* gene duplication and duplication of ventricular chambers in vertebrates [Lin et al., 1997; Cripps and Olson, 2002; Olson, 2006; Wagner, 2008]. In addition to the expansion of the ancestral cardiac gene network, co-opting new genetic networks (*Mesp, Isl1, Foxh1*) is also involved in the evolution of the four-chambered heart [Olson, 2006].

Hox genes encode a family of related transcription factors with similar DNA binding sites. They are most important for patterning on the head-tail body axis during development of animal embryos as well as formation of body plan and structure, and participated in morphological changes during animal evolution. *Hox* genes went through cycles of duplication during evolution. The diploblastic placozoan *Trichoplax*, the most basal metazoan which does not possess any defined main body axis, may have only a single *Hox* gene rather than a set of *Hox* genes [Schierwater and Kuhn, 1998]. Cnidarian *Nematostella vectensis* has a seven-gene cluster of three *Hox* genes and four *Hox*-related genes and a separate cluster of two *ParaHox* genes [Ryan et al., 2007]. *Hox* genes of triploblastic animals might have originated from an ancestral *Hox* cluster consisting of three *Hox* genes separated early in metazoan evolution [Schubert et al., 1993]. The last common bilaterian ancestor had a collinear cluster of at least eight *Hox* genes [de Rosa et al., 1999; Garcia-Fernandez, 2005]. The cephalochordate Amphioxus has a single *Hox* cluster with 14 colinear *Hox* genes, the "prototypical" *Hox* cluster [Minquillon et al., 2005]. Mammalians and many other vertebrate genomes have four *Hox* gene clusters which in mice and humans, for example, contain 39 *Hox* genes. The teleost fishes have seven partial *Hox* clusters which contain up to 48 *Hox* genes [Hurley et al., 2005; Lemons and McGinnis, 2006; Duboule, 2007]. Vertebrates also have the most highly organized *Hox* gene clusters [Duboule, 2007]. The first cycle of duplication of *Hox* genes occurred during the transition between cephalochordates and jawless vertebrates, the second during transition between jawless and jawed vertebrates. Fish-specific *Hox* cluster duplication is coincident with the origin of teleosts

[Crow et al., 2006]. The whole period was characterized by the highest rate of increase of morphological complexity [Holland, 1996; Holland and Garcia-Fernandez, 1996; Aburomia et al., 2003; Duboule, 2007; Khaner, 2007; De Robertis, 2008]. Cluster duplications assisted the progress of global regulations through involvement of globally acting enhancers (reviewed in [Duboule, 2007]). Thus the greater number of *Hox* genes and their clusters, the increase in the general organization of the clusters and the global regulation enhancement were important for the growth of complexity and adaptive evolution of vertebrates [Crow et al., 2006; Duboule, 2007].

The *Hox* gene cluster, together with *ParaHox* and *NK* gene clusters, comprise the *ANTP* megacluster. *Hox* genes are expressed predominantly in the neuroectoderm, *ParaHox* genes − in endodermal derivatives and *NK* genes − mostly in mesodermal derivatives. It was hypothesized that appearance of the megacluster of homeobox genes determined the appearance of the three germ layers in metazoans [Garcia-Fernandez, 2005].

Similarly to animal *Hox* genes, in plants the family of MADS box transcription factors regulates developmental patterning. The network of MADS transcription factors is expressed in the developing flower and participates in the regulation of a variety of flower developmental processes. Evolution of flowering plants (angiosperms), the most successful group of land plants, is accompanied by the extensive duplication of MADS box genes. Green algae have only one MADS box gene, the most recent common ancestor of gymnosperms and angiosperms may have had seven, the angiosperm *Arabidopsis thaliana* has 107, and rice, another widely diverged angiosperm species, has at least 71 (reviewed in [Irish and Litt, 2005; Wagner, 2008]).

Whole-genome duplications (WGDs) may be connected to an increase in morphological complexity. The two rounds of genome duplication in ancestors of vertebrates were followed by important morphological innovations such as enhanced nervous, endocrine and circulatory systems and sensory organs; more complex brains; the skull, vertebrae, endoskeleton and teeth; and in jawed vertebrates, paired appendages, hinged jaws and an adaptive immune system (reviewed in [Van de Peer et al., 2009]). The authors conclude that WGD events, although rarely on evolutionary scale, led to an increase in biological complexity and the origin of evolutionary novelties [Van de Peer et al., 2009].

Developmental constraint at the level of gene duplication was discovered in *Caenorhabditis elegans* and *Caenorhabditis briggsae*. Genes expressed after embryogenesis have a significantly greater number of duplicates as compared to early-expressed genes. This may be important for macroevolutionary change and the origin of innovations [Castillo-Davis and Hartl, 2002].

Correlation of phylogenetic differences at later ontogenetic stages with the expression of newly evolved genes supports the role of novel genes in the origin of evolutionary innovations and morphological novelties [Domazet-Loso and Tautz, 2010b]. In *Drosophila*, two times more orphan genes are expressed in adults as compared to embryos [Domazet-Loso and Tautz, 2003]. Gene signatures of the origin of the germ layers were detected on a *Drosophila* phylostratigraphic map arranged in ectoderm-endoderm-mesoderm stepwise progression, with peaks in Opisthokonta, Metazoa and Bilateria phylostratums, respectively [Domazet-Loso et al., 2007]. Orphan genes participate in the origin of morphological innovations and other species-specific novelties in *Hydra* [Khalturin et al., 2009]. Vertebrate-specific genes are more specialized, i.e. characterized by a narrower breadth of tissue expression as compared to pre-Cambrian genes [Subramanian and Kumar, 2004]. Post-Cambrian genes of *Drosophila melanogaster* and zebra-fish are much more likely to have a stage-specific expression than older genes [Tautz and Domazet-Loso, 2011]. Based on these and similar data, a model of punctuated evolution of protein families was suggested which postulates that protein families are initiated by founder genes that represent evolutionary novelties, i.e. new functional proteins or protein domains. Genomic phylostratigraphy was suggested as an approach to study this model [Domazet-Loso and Tautz, 2003, 2008, 2010a; Domazet-Loso et al., 2007]. This approach has shown that the peaks of gene emergence slightly predate the radiations, so that novel genes participate in the origin of evolutionary novelties (reviewed in [Tautz and Domazet-Loso, 2011]).

The data on human gene expression obtained by massively parallel signature sequencing confirm that differences between cell and tissue types are largely determined by transcripts derived from a limited number of tissue-specific genes, rather than by combinations of more promiscuously expressed genes. Most gene products participate in the maintenance of specialized functions [Jongeneel et al., 2005]. Human-specific duplicates evolving under adaptive natural selection include genes involved in neuronal and cognitive functions [Han et al., 2009].

The above discussion demonstrates that there is a connection between the origin of evolutionarily new genes and the emergence of evolutionary innovations and morphological novelties in multicellular organisms.

However, the alternative view exists which claims that the evolution of morphological diversity is connected mainly with the combinatorial changes of pre-existing genes, i.e. new combinations of gene expression. The so-called genetic theory of morphological evolution suggests that morphological form evolves by altering the expression of functionally conserved proteins and that such changes mainly occur in *cis*-regulatory elements (CRE) of developmental regulatory and target genes [King and Wilson, 1975; Carroll, 1995, 2001, 2005, 2008; Davidson, 2006; Prud'homme et al., 2007]. In particular, it proposes that mutations in CREs of existing genes are more common and a source of morphological variation compared with the evolution of new genes [Carroll, 1995, 2008], while the evolution of coding sequences plays a major role in physiological differences between the species [Carroll, 2005].

This theory was originally based on an observed small difference between the proteins of chimps and humans and a small (about 1.5%) divergence between human and chimp DNA [King and Wilson, 1975; Chen and Li, 2001].

This theory was criticized by many authors. Roy Britten has shown that divergence between samples of chimpanzee and human DNA sequences is 5%, counting indels. He called the widespread quotations that humans had a 98.5% gene similarity with chimps a mistake [Britten, 2002]. Matthew Hahn and co-authors have shown that 6.4% of all human genes (1418/22,000) do not have a one-to-one ortholog in the chimpanzee. A disproportional amount of gene gain and loss has occurred between humans and chimpanzees: since their split 5–6 million years ago 678 genes have been gained in the human genome and 740 genes have been lost in the chimp genome [Demuth et al., 2006; Hahn et al., 2007]. Studies of lineage-specific gene duplication and loss in humans and several great apes discovered more genes with copy number expansions in humans, including brain-related genes [Fortna et al., 2004]. These and many other studies raise the possibility that lineage-specific expansion and contraction of gene families may play a role in evolution comparable to that of changes in orthologous sequences; this may help to explain the great morphological differences between humans and chimpanzees [Hahn et al., 2007; Demuth and Hahn, 2009].

Data on positive selection in human regulatory regions have been published [Hahn, 2007; Haygood et al., 2007].

Hopi Hoekstra and Jerry Coyne argued that there is no theoretical or empirical basis for the contention that morphological and functional adaptations evolve by different genetic mechanisms. They presented evidence that both genomic and single-locus studies detected many adaptive mutations in protein-coding regions, while examples of adaptive cis-regulatory mutations are fewer and focus on trait loss rather than gain [Hoekstra and Coyne, 2007].

Sean Carroll responded that the latter critics conflated the evolution of animal form and adaptation [Carroll, 2008].

The discussion of these fundamental issues will certainly continue and may eventually result in the synthesis of two approaches that are not mutually exclusive. Evolution may proceed through a combination of structural and cis-regulatory mutations [Hoekstra and Coyne, 2007; Hoekstra, 2012]. When mutations in regulatory and coding sequences were studied in the model of adaptive evolution in three-spined sticklebacks, the whole-genome sequences provided information that both coding and regulatory changes occur in loci associated with marine–freshwater evolution [Jones et al., 2012]. Alternative regulation levels may also be as effective as enhancer sites in producing mutational targets for developmental evolution [Alonso and Wilkins, 2005].

The genetic theory of morphological evolution and the concept of evolution by gene duplication do not necessarily contradict each other. At least some novel genes may be regulatory, and structural mutations may happen in genes coding for transcription factors. For example, in *Hydra* both structural and regulatory orphan genes may participate in building cnidocytes — the "stinging cells," one of the most complex cell types in multicellular organisms [Khalturin et al., 2009]. Both pre-existing genes [Donoghue et al., 2008] and novel genes [Martinez-Morales et al., 2007] participated in the evolution of the neural crest differentiation program. Co-option, gene duplication and alternative splicing participated in the origin of the vertebrate brain from the invertebrate chordate brain. As a result, vertebrates have a greater number of proteins that are involved in new brain functions [Holland and Short, 2008].

After the first genomes were sequenced it became evident that the gross number of protein-coding genes in the genomes does not directly correspond to the complexity of the organisms, e.g. if we compare the genomes of a fly, a nematode and human. This discrepancy was termed G(gene)-value paradox [Betran and Long, 2002; Hahn and Wray, 2002], by the analogy with C-value paradox terminology used before to describe the lack of correlation of genome size and complexity of organisms.

There could be several reasons for the G-value paradox. The first one may reside in underestimation of the gene number in humans. For example, alternative splicing generates at least 78,000 different mammalian proteins [Carninci, 2006]. The estimate of the number of genes for noncoding RNAs and functional pseudogenes has just started; their number is considerable and should be taken into account. The number of long ncRNAs exceeds 23,000 transcription units in mice [Katayama et al., 2005], which is more than the different protein coding RNAs, and similar complexity is revealed for human ncRNAs [Carninci, 2010]. The available data from analysis of sequenced genomes suggest that the relative amount of noncoding DNA increases with the increase of complexity [Taft et al., 2007].

Secondly, amphioxus and tunicates are more appropriate to compare with vertebrates, not flies and worms. As I discussed above, the number of *Hox* genes, which are connected with the evolution of complexity, is increasing in vertebrates. The same is true for *ParaHox* genes and many other genes and gene families due to massive gene duplication on the vertebrate lineage [Holland, 2003].

Finally, one of the reasons for the G-value paradox could be that not all the genes in the genomes of multicellular organisms are directly connected with the evolution of complexity, e.g. with the number of terminally differentiated cell types.

Population geneticists claim that the increase of genome complexity in multicellular organisms as compared to prokaryotes may have evolved passively as a consequence of reduction of populations' size and the increase in organisms' size. That is, a considerable part of the genomic complexity of multicellular organisms may not have arisen as a response to selection for new cell types and functions. Nevertheless, the secondary use of increased genomic complexity in adaptive phenotypic evolution is possible [Lynch and Conery, 2003].

9.2.1 Where are the Novel Genes Expressed and Which Cells Give Rise to Evolutionary Innovations and Morphological Novelties?

Scientists who study the origin of novel genes usually do not ask the question of where these genes will be expressed. The preconceived idea is that there are enough conditions for the new genes to be expressed in pre-existing cells, or that extra cells will be automatically generated in organisms in adequate amounts for the expression of new genes and their novel functions. Nevertheless, many authors since August Weismann [Weismann, 1893] have pointed out the existence of competitive relationships and relationships of incompatibility between the genes which will disturb such a scenario (reviewed in [Kozlov, 1979]). The expression of novel genes in the same cell may have energy constraints and selective disadvantages [Wagner, 2005].

Studies of the origin of evolutionary innovations and morphological novelties raise the question, "Due to what processes are the extra cell masses necessary for such innovations and novelties generated?" It was suggested that undifferentiated "set-aside" cells with indefinite division potential may be the substrate for the origin of large structures in bilaterians [Davidson et al., 1995]. To satisfy such a goal the set-aside cells should be disconnected from cell division and differentiation controls imposed by Type 1 embryogenesis mechanisms. Eric H. Davidson and co-authors thoughtfully wrote, "Thus genetic misexpression of various oncogenes may result in release from cell replication controls and loss of differentiated phenotypes, as in retroviral tissue culture transformation. Whatever their origin, the evolutionary 'invention' of developmental programs for the generation of set-aside cells was among the primary causes of the appearance of the higher Metazoa" [Davidson et al., 1995]. Although criticized (see Parts 10.9 and 10.10 for more discussion of the concept of "set-aside" cells), this view is also supported by analysis of *Hox* genes role in development of modern bilaterians, which suggests that the evolution of *Hox* genes could indeed occur in concert with the set-aside progenitor cells in stem bilaterians [Arenas-Mena et al., 1998; Peterson et al., 2000a,b; reviewed in Knoll and Carroll, 1999]. I would point out that cell populations with indefinite division potential and loss of cell division and differentiation control resemble tumors as they are usually defined. Intuitively, the concept of set-aside cells is reminiscent of the concept of embryonic remnants discussed above in Part 5.1.

The extra cells masses were needed not only during the appearance of bilaterians, but also for the origin of latter morphological novelties. For example, the origin of the right ventricle of the vertebrate heart needed, besides duplication of the *Hand* gene, the recruitment of a novel population of precursor cells instead of the simple expansion of pre-existing precursor cells [Olson, 2006]. There are two heart fields with progenitor cells that participate in building the mammalian heart [Buckingham et al., 2005]. The transforming Growth Factor β (TGFβ)-Smad signaling pathway specifies the anterior heart field which forms the right ventricle of the heart and the outflow tract where specialized expression of *Hand 2* and *Mef2c* takes place [von Both et al., 2004]. Interestingly, TGF-β mediates tumor suppression in normal cells and facilitates cancer progression in malignant cells [Tian et al., 2011].

The question was raised of whether or not gene duplication is somehow connected with morphological duplication. Although morphological duplication has some common features with gene duplication, especially in subfunctionalization and increased functional complexity, there is no evidence that these processes are directly connected. Experiments on genetically induced morphological duplication have shown functional redundancy of duplicates. Such redundancy could provide a substrate for subfunctionalization and neofunctionalization. But the major question remains about the origin of the cellular material used to generate the duplicates (reviewed in [Hurley et al., 2005]).

Scientists who assume that there is a general correlation between the increase in the gene number in the genomes of evolving organisms, from one side, and the increase in the number of cell types, the origin of other innovations and the overall complexity, on the other, should explain how such adequate correlation was realized at the cellular level. An adequate increase in cell number that accompanied the process of the origin of novel genes is hard to imagine. More likely, some autonomous cellular proliferative processes were recruited to provide the space for the expression of new genes.

9.2.2 Similarly to Gene Duplication and Retrotransposition Events, Tumors May Provide "Raw Material" for the Evolution of Complexity

Gene duplication provides "raw material" for the mutational process. The "extra" gene copies, functionally unnecessary to the organism, are used in genome evolution and the origin of novel genes. In a similar way, tumors may provide "extra" cell masses, functionally unnecessary to the organism, which could be used in progressive evolution for the origin of evolutionary innovations and morphological novelties.

Although pathological to individual organisms, tumors still may play an evolutionary role. The analogy of tumors with mutational process in this respect has already been discussed above. Here I will refer to the analogy with the role of non-LTR retrotransposons in genome evolution. As discussed above, non-LTR retrotransposons LINE-1, *Alu* and SVA played an important role in human genome evolution. Haig Kazazian called mobile elements the "drivers of genome evolution" [Kazazian, 2004]. At the same time, they cause many human genetic diseases by *de novo* insertions [Kazazian et al., 1988; Deininger and Batzer, 1999; Chen et al., 2005; Callinan and Batzer, 2006; Belancio et al., 2008]. The estimated retrotransposition rates of *Alu*, LINE-1 and SVA are very high — one in 21 births, 212 births and 916 births, respectively [Xing et al., 2009].

Similar to gene duplication and retrotransposons, tumors might provide the raw material — extra cell masses — for the evolution of multicellular organisms, i.e. they could be "drivers" of the evolution of multicellularity just as mobile elements were the drivers of genome evolution.

Chapter 10

The Origin of New Cell Types, Tissues and Organs by Tumor Neofunctionalization

10.1 THE HYPOTHESIS OF EVOLUTION BY TUMOR NEOFUNCTIONALIZATION

Since the 1970s, I have been interested in the problem of the evolutionary role of tumors in the context of the interrelationship between processes occurring at molecular, cellular and multicellular levels. The result was the development of the hypothesis of evolution by tumor neofunctionalization [Kozlov, 1979, 1983, 1987, 1988, 1996, 2008, 2010]. The hypothesis is formulated as follows:

Tumors are the source of extra cell masses which may be used in the evolution of multicellular organisms for the expression of evolutionarily novel genes, for the origin of new differentiated cell types with novel functions and for building new structures which constitute evolutionary innovations and morphological novelties.

Hereditary tumors may play an evolutionary role by providing conditions (space and resources) for the expression of genes newly evolving in the DNA of germ cells. As a result of expression of novel genes, tumor cells may acquire new functions and differentiate in new directions which may lead to the origin of new cell types, tissues and organs. The new cell type is inherited in progeny generations due to genetic and epigenetic mechanisms similar to those for pre-existing cell types (Figure 10.1).

Tumors at the early stages of progression, benign tumors, pseudoneoplasms and tumor-like processes which provide evolving multicellular organisms with extra cell masses functionally unnecessary to the organism, are considered as potentially evolutionarily meaningful. Malignant tumors at the late stages of progression, however, are not.

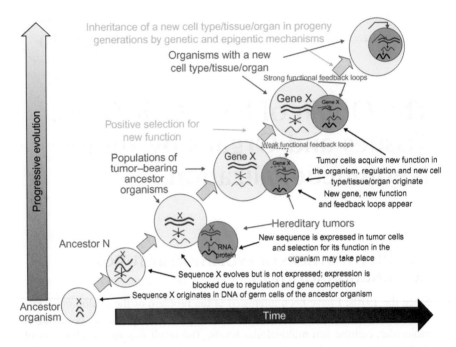

FIGURE 10.1   Diagram explaining the hypothesis of evolution by tumor neofunctionalization. *Modified from [Kozlov, 2010], with permission.*

10.2   GENE COMPETITION AND THE POSSIBLE EVOLUTIONARY ROLE OF TUMORS

Originally the idea of the possible evolutionary role of tumors was prompted to the present author by the concept of gene competition [Kozlov, 1976, 1979, 1983, 1987, 1988, 1996]. The concept of gene competition was coined simultaneously with the concept of the gene. August Weismann was the first to introduce this concept in his theory of the germ plasm [Weismann, 1893]. C.H. Waddington [Waddington, 1948] and S. Spiegelman [Spiegelman, 1948] considered gene competition as an immediate cause of cell differentiation. But after the era of regulation came, gene competition as a general principle was partially forgotten although the number of examples of competitive relations between macromolecules was increasing [Kozlov, 1979]. The concept of gene competition does not contradict the concept of gene regulation. Gene competition appeared during evolution before gene regulation, and differential expression of genetic information might have evolved from competitive relations between the genes [Kozlov, 1976, 1979]. Thus gene competition may constitute a more general principle as compared to gene regulation. The principle of gene competition could be formulated as follows: at the different levels of the expression of genetic information genes (via gene products) come into relations of struggle for common resources and space, the consequence of which is mutual reduction of their activity [Kozlov, 1976, 1979]. With an increase of gene number in the genomes of evolving multicellular organisms, the intensification of gene competition and the appearance (in some cases) of antagonistic relations between the genes should take place. The consequence of antagonistic relations between the genes, or relations of incompatibility between the genes, is complete repression of certain genes' expression. I called antagonistic or incompatible those genes which under certain circumstances come into antagonistic relations [Kozlov, 1979, 1996].

Because of gene competition and incompatibility between the genes, the pre-existing cell types of multicellular organisms had restricted potential for the expression of newly evolving genes. The origin of novel genes during genome evolution would lead to an increase in gene competition and incompatibility between the genes at the level of their expression during the normal development. Because of gene competition, some of the newly evolving sequences may stay silent and never express their genetic information during ontogenesis [Kozlov, 1979, 1996] (Figure 10.1).

Therefore, multicellular organisms would need extra cell masses functionally unnecessary to the organism for the expression of newly evolving genes. The pre-existing cell types cannot provide such excess cell masses because of regulation and limitations imposed on the number of possible cell divisions, and also because of their functional necessity to the organism. Tumors could provide the evolving multicellular organisms with the extra cell masses functionally unnecessary to the organism for the expression of newly evolving genes [Kozlov, 1979, 1996].

I suggest that tumors could be a sort of proving ground (or reservoir) for the expression of newly evolving genes that originate in the course of genome evolution in the DNA of germ cells (i.e. not in tumor cells themselves) [Kozlov, 2010]. When the expression of an evolving sequence in tumors coincides with the origin of a new function beneficial to the organism, some tumor cells would differentiate in a new direction, resulting in a new cell type for the given multicellular species (Figure 10.1). The differentiation in new directions is feasible in tumor cells because of their greater plasticity as compared to normal cells and their capability to differentiate and also because unusual genes and gene sets, including novel genes, are activated in tumors, as discussed in Chapter 4.

Using similar terminology, other authors [Summers et al., 2002; Crespi and Summers, 2005, 2006] suggested that there is a link between intragenomic conflict and cancer. Summers and co-authors refer to A. Pomiankowski's definition of intragenomic conflict: "Intragenomic conflict occurs because there are multiple genetic entities within the genome (e.g. paternal and maternal halves of the nuclear genome, cytoplasmic element genomes, transposable elements, etc.), and these elements do not all have the same mode of inheritance, the same opportunities for self-replication, the same relatedness to genetic elements in other genomes, or the same information about their relatedness to other genetic elements in other genomes" [Pomiankowski, 1999]. After reviewing genomic imprinting and growth factors, mammalian X and Y chromosomes' evolutionary "arms race," mitochondrial genes and transposable genetic elements including retroviruses, Summers and co-authors conclude that "by their very nature, genes involved in intragenomic conflict are likely to be disruptive and to promote uncontrolled growth and replication. Such genes are inherently likely to be exploited by cancer cell lineages as they evolve toward escape from control by the body" [Summers et al., 2002].

In another paper, Bernard Crespi and Kyle Summers hypothesize that antagonistic coevolution establishes links between positive selection at the molecular level and increased cancer risk [Crespi and Summers, 2006]. According to Crespi and Summers, antagonistic coevolution includes evolutionary conflicts between males and females, mothers and fetuses, hosts and parasites, and other parties with divergent fitness interests. The focal point for antagonistic coevolution is resource acquisition and use. I quote, "Thus, strong selection in the context of antagonistic coevolution leads to evolutionary change in genes and traits related to conflict, and the pleiotropic effects of these changes may generate increased cancer risk" [Crespi and Summers, 2006]. Nevertheless, other authors point out that the exact nature of selective pressures caused by intragenomic conflict is not well understood [Zhang and Roseberg, 2002].

The intragenomic conflict and antagonistic coevolution may be considered as a particular case of gene competition as defined in [Kozlov, 1979, 1996]. The definition of intragenomic conflict is narrower as compared to the principle of gene competition, although at the end it comes down to a struggle for resources. I infer that antagonism between the genes is neutralized by spacio-temporal disconnection of their products, which leads, in coincidence with relatively independent tumor growths, to the origin of new cell type/tissues/organs in the evolution of multicellular organisms. Summers, Crespi and co-authors suggest that antagonistic coevolution generates an increased cancer risk, although the mechanisms of this linkage are not well understood.

10.3 THE POSSIBLE EVOLUTIONARY ROLE OF CELLULAR ONCOGENES

In light of the hypothesis of evolution by tumor neofunctionalization, the evolutionary role of cellular oncogenes might consist in sustaining a definite level of autonomous proliferative processes in evolving populations of multicellular organisms and in promoting the expression of evolutionarily new genes. After the origin of a new cell type/tissue/organ, the corresponding oncogene (together with its suppressor genes) should have turned into a cell type-specific regulator of cell division and gene expression [Kozlov, 1996]. The current state of knowledge in this area (e.g. the *Xiphophorus* model data) supports such an interpretation.

The widespread occurrence of cellular oncogenes among different groups of multicellular organisms [Bishop, 1983], the evolutionary conservatism of cellular oncogenes and tumor suppressor genes [Shilo and Weinberg, 1981; Atchley and Fitch, 1995; Thomas et al., 2003], the specificity of expression of cellular oncogenes [Bishop, 1983; Muller and Verma, 1984], their ability to activate other cellular genes [Kingston et al., 1985; Weinberg, 1985] and the role of cellular oncogenes in embryogenesis (see Part 5.1) are in favor of their positive evolutionary role. An increase in the intensity of purifying selection exerted on oncogenes and tumor suppressor genes relative to that exerted on other genes has also been found [Thomas et al., 2003]; i.e. these genes are in a certain sense more important to the organisms than other genes.

10.4 THE ORIGIN OF FEEDBACK LOOPS REGULATING NEW FUNCTIONS, NEW GENE EXPRESSION AND NEW CELL TYPE PROLIFERATION

The origin of the new function and the new gene should be connected with the new regulatory feedback loops, as demanded by the general theory of regulation [Rosenblueth et al., 1943]. I believe that if the origin of new organismal functions in evolution is connected with tumors, it is new functional feedback loops that may eventually convert the tumor into a new cell type/tissue/organ.

As I discussed earlier in Part 7.5, the placenta is a tumor-like organ which differs from a tumor because it is strongly regulated by the host. Placental extravillous trophoblast (EVT) cells are under powerful control of many growth factors, growth factor-binding proteins, proteoglycans and extracellular matrix components. Transforming growth factor (TGF)-β produced by maternal decidua is the main negative regulator of EVT cells [Lala et al., 2002].

A similar feedback loop was suggested to describe the interrelationship of local cell proliferation, *Hox* gene expression and morphogenesis [Duboule, 1995]. According to Denis Duboule, the cross-regulatory interactions in the form of a feedback loop control the balance between growth and patterning.

When the sequence in DNA of germ cells acquires organismal function, it becomes a gene. It may be a protein-coding gene or a gene for noncoding RNA. Selection of organisms with a new function would tend to enhance this function as far as it is beneficial to the organism. Selection for enhancement of the function of novel gene(s) is connected with a higher amino acid substitution rate in evolving proteins ($Ka/Ks > 1$) in progeny organisms. This is so-called positive selection. Examples of the positive selection of novel genes were discussed in Part 9.1.1.

Natural selection of organisms with the novel gene/function/cell type should also strengthen new feedback loops that regulate the new function. In an analogous way, the "partner fidelity feedback" is strengthened in the evolution of mutual benefits across multiple generations of symbiotic partners [Dethlefsen et al., 2007].

Feedback loops on both the molecular and cellular levels are needed to regulate the new function. The feedback loop at the molecular level is necessary to regulate the new gene expression and would include transcriptional factors, small RNAs and other molecules. The feedback loop at the cellular level is necessary to regulate proliferation of stem/progenitor cells of the new cell type and would include growth factors, interleukins, extracellular matrices and T cells. At the earlier stages of the origin of the new function, a host immune response may play a role in supporting a tumor in an equilibrium state (Figures 10.1–10.3).

Cancer/testis (CT) antigen genes, which may be considered as evolutionarily novel genes without any known function for many of them as yet (see Part 11.1.4), demonstrate both types of feedback. The feedback loop at the molecular level is the autoregulatory loop which includes p53 and CT antigens. The CT antigen regulatory loop includes activating and repressive

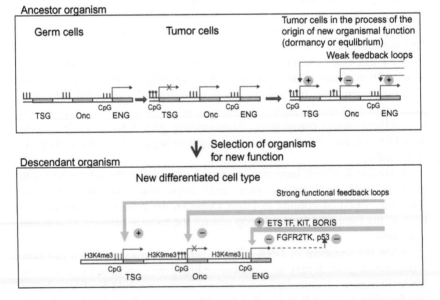

FIGURE 10.2 The origin and evolutionary enhancement of functional molecular feedback loops in the new cell type. TSG – tumor suppressor gene; Onc – cellular oncogene; ENG – evolutionarily novel gene.

signals. Signals that activate the expression of CT antigen genes are ETS transcription factors, a receptor tyrosine kinase protein KIT, the BORIS protein, which is a paralog of the imprinting regulator and chromatin insulator protein CTCF, and hypomethylation of CpG-rich promoters of CT antigen genes. Repressive signals include FGF receptor (FGFR)2 tyrosine kinase activation and p53. In their turn, CT antigens inhibit p53 function. Such regulatory interactions result in equilibrium of expression of CT antigen genes in tumors [Akers et al., 2010] (Figure 10.2).

In addition, CT antigens participate in immunological control of CT antigen—expressing cells by CT antigen—directed CTLs, which bring to the origins of cellular feedback loops controlling proliferation of cells with a newly evolving function (Figure 10.3).

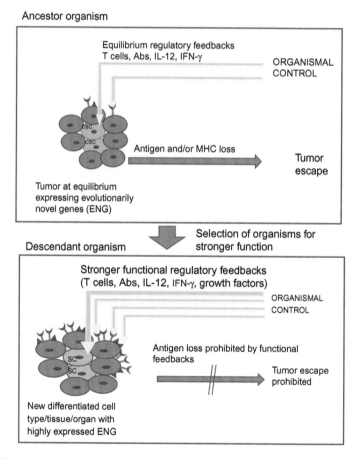

FIGURE 10.3   The origin and evolutionary enhancement of functional regulatory cellular feedback loops. CSC — cancer stem cell; SC — normal stem cell.

More examples of cellular feedback loops that may play a role in regulation of novel functions are provided by the phenomena of tumor dormancy
and equilibrium. A dormant tumor consists of latent tumor cells that reside
in the organism for decades before eventually resuming growth as either
recurrent primary tumors or metastases. Tumor cell growth arrest (tumor cell
dormancy) and control of the expansion of a dividing tumor cell population
(tumor mass dormancy) contribute to tumor dormancy [Aguirre-Ghiso,
2007]. Equilibrium is a stage in the process called tumor immunoediting in
which the growth of occult tumors is controlled by immunity. It may represent a type of tumor dormancy [Schreiber et al., 2011].

Tumor cell dormancy may be due to a G0/G1 arrest or differentiation as
a result of reduction of crosstalk with the tumor cell microenvironment. For
instance, inhibition of β1-integrins by integrin-blocking antibodies leads to
the differentiation of breast cancer cells into non-proliferating acinar structures [Weaver et al., 1997].

Tumor mass dormancy may be a consequence of immune response.
CD4$^+$ and CD8$^+$ T cells, IL-12 and IFN-γ are responsible for keeping tumor
cells dormant. Equilibrium is a function of adaptive immunity only. Both the
growth inhibitory and cytocidal actions of immunity on tumor cells lead to
equilibrium [Schreiber et al., 2011] (Figure 10.3).

B cell lymphoma becomes dormant after interaction with antibodies
against the idiotype on its immunoglobulin receptor. Anti-idiotypic immunity
does not eradicate BCL-1 tumor cells but maintains them dormant over an
extended period. Antibody alone is enough to induce the dormant state. The
majority of the dormant cells are arrested in G0/G1. The authors conclude
that antibodies interacting with the Ig receptor may act as an agonistic ligand
that induces growth inhibitory signals in the cells [Rabinovsky et al., 2007].

Monoclonal antibodies to the HER-2/Neu receptor either inhibit or accel-
erate the tumor growth of HER-2/*neu* transfectants in athymic mice. Tumor-
inhibitory antibodies induced cellular differentiation of various cultured
human breast cancer cells. The extent of differentiation is correlated with the
tumor-inhibitory potential of different antibodies. On the other hand, tumor-
stimulatory antibodies do not cause differentiation [Bacus et al., 1992].

A tumor-specific immune response is regulated by tumor antigen levels.
Loss of tumor antigen expression and MHC leads to tumor escape from
immunological control and promotes tumor growth. On the contrary,
increased tumor antigen expression is necessary for immunological surveillance and overcomes immunological "ignorance" to solid tumors [Spiotto
et al., 2002; Kim et al., 2007; Schreiber et al., 2011]. That is why we may
infer that selection for the enhancement of the novel function, which leads
to the increase of the expression of the novel gene and novel antigen, will
also lead to the increase of immune response and stabilization of the tumor
at the equilibrium stage, in its transition to acquiring its new function in
the organism.

More tumor cells would terminally differentiate in a new direction as a consequence of selection for enhancement of the new function and the increase of expression of the novel gene. Tumor stem cells would acquire developmental and homeostatic control mechanisms, determined by the new function, and give rise to normal stem cells of the new cell type/tissue/organ. The feedback loop that interconnects stem cell proliferation and differentiation will appear and evolve in parallel with the evolution of new function. It will be strengthened in parallel with the enhancement of the new function in generations of organisms, and eventually the tumor stem cells will turn into normal stem cells of the new type, and new differentiated cell types/tissues/organs will appear marking a new stage in progressive evolution. Diagrams presented in Figures 10.1–10.3 summarize this consideration.

Such a scenario is supported by many facts. Outgrowths derived from dormant *Wnt1*-initiated mammary cancer can participate in reconstituting morphologically normal ductal trees of functional mammary glands [Gestl et al., 2007]. Half of the genes appearing first in vertebrates encode growth factors with a role in committing precursor fate for the neural crest derivatives. The same is true for other vertebrate-specific tissues. For example, vertebrate-specific interleukins and hematopoietic cytokines regulate lymphoid and blood cell lineages [Martinez-Morales et al., 2007]. Evolution of the growth factor genes may be a part of the functional feedback loop discussed above.

The new gene may be incorporated into existing but modified regulatory circuits. The molecular regulation may co-opt pre-existing regulatory mechanisms, such as the transcription factor network and evolutionary conserved pathways, as in the case of hematopoietic cells [Hoang, 2004]. Recruitment of regulatory genes or pre-existing regulatory networks may be common, but the acquisition of new transcription factors and/or additional regulatory elements might be necessary, as in the case of cephalopod morphological novelties [Lee et al., 2003]. The vertebrate- and mammalian-specific cytokines, also interacting with the ancient Jak-STAT pathway, regulate progenitors with restricted potential and mature functional cells [Hoang, 2004]. The growth factors, homeobox transcription factors, cell surface components and extracellular matrix interact in the process of cell type genesis and organogenesis [Cillo et al., 2001].

The epigenetic progenitor model of the origin of tumors suggests that perturbation of the normal balance between undifferentiated progenitor cells and differentiated committed cells could be one of the earliest steps in neoplastic transformation. The expansion of progenitor cells may be caused by epigenetic abnormalities of "tumor-progenitor genes" that normally regulate stemness and constitute parts of an epigenetic network with positive and negative feedbacks [Feinberg et al., 2005]. Such epigenetic abnormalities are reversible, as evidence on tumor cell differentiation proves.

I suggest that epigenetically caused tumor-related abnormalities in differentiation are also reversible towards the new directions of differentiation which are determined by the expression of evolutionarily new genes in expanded populations of progenitor cells. Thus the epigenetic expansion of progenitor cells would be put under control by establishing new developmental and homeostatic control mechanisms and new feedbacks in the pre-existing epigenetic network (Figures 10.1–10.3).

A continuum model for tumor suppression suggests that tumor suppression is related to a continuum of tumor suppressor gene (TSG) expression. A continuum of increasing TSG expression will be negatively correlated with malignancy, whereas increasing oncogene expression will be positively correlated with malignancy [Berger et al., 2011]. Evolution of tumor stem cells' developmental and homeostatic control mechanisms in parallel with the emergence of new functions discussed above may also include the gradual enhancement of the activity of corresponding TSG and the decrease of activity of cellular oncogenes (Figure 10.2). This would lead not only to the continuum mode of tumor suppression in progeny organisms, but to specific regulation of novel tissues, as exemplified above by data on the connection of protooncogenes and tumor suppressor genes with normal development (see Part 5.1), and is also supported by tissue specificity and context dependency of tumor suppression [Berger et al., 2011].

10.5 HOW IS THE NEW CELL TYPE INHERITED IN PROGENY GENERATIONS?

First of all, as I noted above, the novel cell type-specific genes originate in the DNA of germ cells and are inherited in generations of organisms. Heritability of tumors in which evolutionarily novel genes are expressed provides the cellular basis for heritability of the novel cell type. Epigenetic signals and modifications determine the inheritance of gene expression patterns of the novel cell types in the progeny generations. Epigenetic information is transmitted through meiosis and gamete formation in multicellular organisms, giving rise to transgenerational inheritance, despite considerable DNA methylation reprogramming in mammalian germ cells and early embryos [Feng et al., 2010]. In a similar way, the transgenerational inheritance of epigenetic signals may lead to fixation of a new cell type in progeny generations. The new cell type will then be inherited due to genetic and epigenetic mechanisms similar to those for pre-existing cell types of multicellular organism.

Epigenetic signals are responsible for the establishment and maintenance of transcriptional states that determine differentiation of cells in multicellular organisms. Self-propagating transcriptional states that are maintained through feedback loops and networks of transcriptional factors (TFs) are the most common type of *trans* epigenetic states. After each cell division, inherited TFs, including some small RNAs, resume their *trans* function on regulatory DNA sequences. *Cis* epigenetic signals are physically associated and inherited along with the chromosome on which they act — as DNA methylation or as changes in histones. The appearance of multicellularity expanded the repertoire of *cis* epigenetic signals in order to deal with the increased complexity and number of transcriptional networks. Epigenetic states that are encoded in *cis* need to be set only once, and many transcriptional patterns can be maintained by a relatively small number of common molecular pathways. Nevertheless, in some sites DNA methylation may be a consequence of transcription factor binding. The epigenetic modifications can mediate both a short-term (mitotic) and long-term (meiotic) transmission of an active or silent gene state without changing the primary DNA sequences [Saze, 2008; Bonasio et al., 2010; Schubeler, 2012].

Many epigenetic signals appear capable of meiotic transmission, including maternally deposited TFs and piRNAs [Brennecke et al., 2008], RNAs involved in paramutation in mice [Rassoulzadegan et al., 2006], histone modifications in sperm chromatin [Hammoud et al., 2009] and DNA methylation in plants [Saze, 2008; Schmitz et al., 2011]. Moreover, DNA methylation may be influenced by the underlying DNA sequence [Schubeler, 2012].

After selection of a certain number of generations of organisms, the new organismal function will be enhanced, more new normal stem-progenitor cells will terminally differentiate into new cell types and tumors from atypical nonfunctional organs will turn into normal regulated functional organs inherited in progeny generations (Figures 10.1–10.3).

10.6 TUMOR-BEARING ORGANISMS AS EVOLUTIONARY TRANSITIONAL FORMS

Fossil records do not contain chains of morphological intermediates that link present-day phyla with their common ancestors, or phyla with different levels of complexity. Tumor-bearing organisms with genetically or epigenetically programmed tumors could represent such transitional forms between established species at different stages of progressive evolution (Figure 10.4). An example of transitional populations of tumor-bearers could be tumor-bearing voles, discussed above. During certain periods of phylogenesis, differentiation of tumor cells in different organisms of such populations could be frequent enough to result in populations of organisms with a new cell type. The organisms with the new cell type would then be selected for their fitness and competitive abilities. *Xiphophorus* fishes, in which there is sexual selection for the spotted caudal melanin pattern, could be an example [Fernandez and Morris, 2008].

New cell types could participate in the formation of new tissues and organs. It is interesting in this regard that the different species of hadrosaurs discussed above in connection with the higher prevalence of tumors featured peculiar osseous outgrowths on their heads, which were shaped as crests and are hypothesized to have been involved in generating sounds [Alexander,

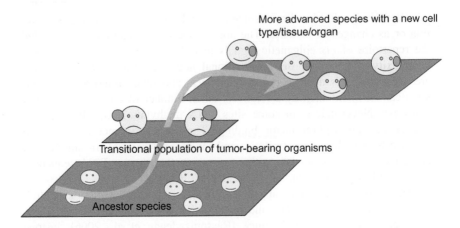

FIGURE 10.4 Population of tumor-bearing organisms with heritable tumors as transition between established species of organisms at different levels of complexity. *From [Kozlov, 2010], with permission.*

2006]. Other examples of new organs originating from tumors could be root nodules, considered as new organs of the plant; the peculiar hood on the head of the goldfish, var. oranda, described by us as a benign tumor; and the placenta, the regulated tumor. These examples were discussed in more detail in Chapter 7.

Unstable transitional forms are known in the evolution of structural elements at different levels of structural organization. For example, unstable heavy elements represent transitional forms in the evolution of transuranium elements. Their stability is determined by a certain ratio between neutrons and protons in the atomic nuclei. In chemistry, unstable molecules with high reactivity are known. In a similar way, tumor-bearing organisms may represent unstable transitional forms between species at different levels of progressive evolution. Their "stabilization" is achieved through the origin of new genes, new functions and functional feedbacks.

10.6.1 Morphological Novelties, Developmental Constraints and Tumors

As defined by Gerd Muller and Gunter Wagner, "[a] morphological novelty is a structure that is neither homologous to any structure in the ancestral species nor homonymous to any other structure of the same organism" [Muller and Wagner, 1991]. These authors point out that morphological novelties have been realized despite developmental constraints that existed in the ancestor organisms. The problem, though, is that the mechanisms through which such transitions happen are not completely understood. The evolution by tumor neofunctionalization hypothesis suggests that tumors may represent a way to overcome developmental constraints. Tumor-bearing organisms may represent transitional forms in the origin of evolutionary innovations. On the other hand, anti-cancer selection may be a source of developmental and evolutionary constraints [Galis and Metz, 2003].

10.7 THE THEORY OF "FROZEN ACCIDENT" MAY BE APPLIED TO THE ORIGIN OF NEW CELL TYPES/TISSUES/ORGANS

The theory of "frozen accident" was suggested to explain the origin of the genetic code. This theory states that the code is universal because any change would be lethal or strongly selected against [Woese, 1967; Crick, 1968]. S. Ohno suggested the view of evolution as a succession of different "frozen accidents" [Ohno, 1973].

The emergence of a new cell type/tissue/organ is a relatively rare evolutionary event associated with progressive evolution and may represent a similar sort of historical (or "frozen") accident. It is hard to imagine that the origin of new genes would directly lead to an adequate increase in the number of cells and cell types in multicellular organisms. Even if new genes are potentially adaptively valuable, they are not by themselves sufficient for the origin of a new cell type. New cell types/tissues/organs may originate as a result of the coincidence of relatively independent processes at different structural levels (i.e. genes, cells and cell masses) and the "freezing" of the valuable coincidences by subsequent selection as in the case of macromelanophores of *Xiphophorus* fishes. Such coincidences may not happen very often. That is why, after a billion years of evolution of multicellular organisms, mammals have acquired only about 200–500 different specialized cell types [Kozlov, 1996; Vickaryous and Hall, 2006] (see also Part 10.10 for further discussion of the possible number of such coincidences in the evolution of vertebrates).

10.8 PROVIDING EXPRESSION OF NEWLY EVOLVING GENES, TUMORS ALSO FACILITATE THE ORIGIN OF NOVEL ORGANISMAL GENE FUNCTIONS

According to the evolution by tumor neofunctionalization hypothesis, tumors provide conditions for the expression of genes and/or sequences newly evolving in the DNA of germ cells. The evolution of such sequences in germ cell DNA and their expression in tumors may result in the emergence of a gene with a potential new function in the organism. Selection in generations of tumor-bearing organisms for this new function may eventually result in the origin of a new function beneficial to the organism and the corresponding novel gene. The important point in this line of reasoning is that evolution at the level of DNA only is not sufficient for the emergence of the new gene and the new function. It must be accompanied by expression and selection for function in the organism of protein and/or RNA products as discussed in Chapter 9. Such expression is possible in tumors, which thus appear not only as a reservoir of expression, but also as a reservoir of the origin of novel organismal gene functions.

Indeed, in tumors that have played a role in evolution, e.g. root nodules of legumes, melanomatous cells of *Xiphophorus* fishes and the tumor-like placenta, evolutionarily novel genes have acquired novel organismal functions: nitrogen fixation in legumes, camouflage and sexual appeal in fishes, and complex placental functions in mammals (see Chapter 7 and Part 12.3 for references).

If a sequence that is expressed in tumors acquires a function beneficial to the organism and becomes an evolutionarily new gene, selection of organisms for enhancement of the new function should take place. This is so called positive selection, and it is connected with a higher amino acid substitution rate in evolving protein ($K_a/K_s > 1$) in progeny organisms. Positive selection of many human tumor-related genes was described in primate lineage [Clark et al., 2003; Pavlicek et al., 2004; Nielsen et al., 2005; Crespi and Summers, 2006; Stevenson et al., 2007; Demuth and Hahn, 2009]. It confirms my prediction concerning expression of evolutionarily new genes in tumors and selection of organisms for new organismal functions of these genes.

The new gene, beneficial to the organism, through involvement into networks of functional feedbacks, would participate in the regulation of tumors in which the gene has been expressed, and in the evolutionary transition of such tumors into the novel functional cell types/tissues/organs. We saw this in the case of the placenta as a regulated tumor (Part 7.5). The *MAGE* gene family of cancer/testis antigens could be another example. The *MAGE* gene family is divided into two subfamilies, *MAGE-I* and *MAGE-II*. The *MAGE-I* subfamily includes chromosome X-clustered genes, *MAGE-A*, *MAGE-B* and *MAGE-C* clusters. *MAGE-I* genes are expressed in the testis and different tumors and code for cancer/testis antigens. On the contrary, *MAGE-II* family genes are expressed in normal tissues. *MAGE-I* genes are of relatively recent origin, and *MAGE-II* genes are relatively more ancient. That is probably why more is known about the functions of *MAGE-II* genes as compared to *MAGE-I* genes. (The same is true for other families of CT genes, X chromosome−located vs. autosomal). For my consideration, it is important that MAGE-II proteins have anti-tumor activity [Cheng et al., 2011; Katsura and Satta, 2011]. This supports my thesis that genes which are expressed in tumors and acquire organismal functions in evolution will participate in tumor regulation and in evolutionary transition of such tumors into the novel functional cell type/tissue/organ.

The origin of new organismal functions in tumors may become a promising target for prevention and treatment of cancers.

10.9 TUMORS AND THE EARLY EVOLUTION OF METAZOA

There are several evolutionary scenarios for the origin of the metazoan body plan which are described in several hypotheses, including the Trochaea hypothesis of Nielsen [Nielsen and Norrevang, 1985; Nielsen, 1995], the benthic colonial hypothesis of Dewell [Dewell, 2000], the larva-like ancestor/set-aside cells hypothesis of Davidson and co-authors [Davidson et al., 1995] and the benthic individual hypothesis [Valentine, 1994; Budd and Jensen, 2000] (reviewed in [Collins and Valentine, 2001; Sly et al., 2003; Malakhov, 2004]).

For my consideration of the evolutionary role of tumors, the set-aside cells hypothesis is the most interesting, although it may not describe the general way of bilaterian origin. My interest is because this hypothesis addresses the problem of the building material for the evolution of bilaterian complexity.

It has been argued that primitive bilaterians were small and had Type 1 embryogenesis which generated the fixed amounts of cells with predetermined fates [Davidson, 1991]. Type I embryogenesis must have preceded the Cambrian radiation. Something should have happened to change this mode of development to bring the variety of organisms with larger bodies and more complex body plans. Eric Davidson and his co-authors believed that ciliated planktotrophic larvae of indirectly developing marine animals provided a model of how that change might have happened in the origin of bilaterians [Davidson et al., 1995; Arenas-Mena et al., 1998; Peterson and Davidson, 2000; Peterson et al., 2000a,b].

Organisms with indirect development have larvae and adult bodies which arise from undifferentiated cells that are set aside from participation in embryogenesis. The larva set-aside cells do not participate in the embryonic specification and differentiation functions, but have functions in the future development. For example, in nematode *Caenorhabditis elegans*, in the first stage larva, of the total 558 cells 42 postembryonic blast cells are set aside from the embryonic functions and their descendants subsequently form additional structures [Sulston et al., 1983; Davidson, 2006]. In ascidian *Ciona* the adult is built largely from larval set-aside cells. In most sea urchins and in other echinoderms the development of the adult body occurs from a specific population of embryonic set-aside cells (so-called maximum indirect development) [Davidson et al., 1995; Davidson, 2006].

It was suggested that undifferentiated set-aside cells with indefinite division potential may be the substrate for the origin of large structures in bilaterians and the larva-like organisms may be the ancestors of modern bilaterians [Davidson et al., 1995]. The authors of the concept wrote that set-aside cells should be disconnected from cell division and differentiation controls imposed by Type 1 embryogenesis mechanisms. They even suggested that misexpression of various oncogenes may participate in the release from cell replication controls and the loss of differentiated phenotypes, and used an analogy with retroviral tissue culture transformation. But they did not go further and call this process a tumor. They wrote that "whatever their origin, the evolutionary 'invention' of developmental programs for the generation of set-aside cells was among the primary causes of the appearance of the higher Metazoa" [Davidson et al., 1995]. I would point out that cell populations with indefinite division potential and loss of cell division and differentiation control resemble tumors as those are usually defined. So tumors or tumor-like processes could be a source of origin of set-aside cells in evolution, in those cases when set-aside cells were formed.

The larva-like ancestor/set-aside cells hypothesis was criticized on comparative biologic and molecular phylogenetic grounds [Conway Morris, 1998; Wolpert, 1999; Jenner, 2000; Strathmann, 2000; Valentine and Collins, 2000; Collins and Valentine, 2001; Sly et al., 2003; Raff, 2008], and, as is now generally accepted, does not represent the actual path of early bilaterian evolution. Nevertheless, the set-aside cells part of this hypothesis is important because it accentuates the necessity of generating the cellular material for the evolution of complexity, and suggests genetic mechanisms which might participate in this process. More traditional zoological theories take for granted that the cellular material will somehow appear.

If the earliest bilaterians developed directly from embryos to adults, as most zoologists currently accept [Sly et al., 2003; Peterson et al., 2005; Raff, 2008], a way around their developmental restriction controls to the genesis of macroscopic animals must have been found before the origin of set-aside cells [Knoll and Carrol, 1999]. That is, cellular proliferative processes based on disconnection from cell division and differentiation controls, which are reminiscent of tumor processes, must have appeared before and independently of set-aside cells. So both scenarios (adult or larval body plans first) demand that tumor-like processes generate extra cell masses for the evolution of macroscopic body plans and large metazoans.

Peterson and co-authors connected processes that delay the specification of cells and thereby allow for growth (tumor-like processes, as I believe) with pattern formation in which *Hox* genes participate [Peterson et al., 2005]. Interestingly, the vertebrate *Hox* genes are involved in relationships between growth and patterning, i.e. in the control of local cell proliferation. The sequential opening of *Hox* gene clusters occurs only in cells with high mitotic activity [Duboule, 1995].

Hox genes are also involved in tumors [Cillo et al., 1999, 2001]. Homeobox genes that are normally expressed in undifferentiated cells are upregulated in tumors, and those that are normally expressed in differentiated tissues are downregulated in tumors [Abate-Shen, 2002]. So it may turn out that the unique role of *Hox* genes in the evolution of bilaterians might include their multiple roles in cell proliferation, differentiation and pattern formation.

Bilaterians arose 630–600 Myr ago. The Cambrian explosion of the diversity of animals started around 530 Myr ago [Peterson et al., 2005]. The diverse complex bilateria appear in the fossil record 544–505 Myr ago, in the Early to Mid-Cambrian, and planktonic larvae emerged about 500 Myr ago, 100 Myr after the origins of basal bilaterian phyla [Raff, 2008].

Paradoxically, it turns out that the origin of tumors should precede and stipulate the origin of macroscopic bilaterians, and could be connected with the Cambrian explosion of metazoan morphological diversity and complexity. Tumor-like processes could be an engine that, together with genome evolution, drove the evolution of bilaterian complexity. It is interesting that genes controlling the programmed cell death, which play an important role in normal development and in surveillance of cancerogenesis, appeared in evolution at about the same period. Robert Horvitz, in his work with the nematode worm *Caenorhabditis elegans*, described three genes involved in programmed cell death – *ced-3*, *ced-4* and *ced-9* genes [Hengartner et al., 1992; Yuan and Horvitz, 1992; Yuan et al., 1993]. These genes are evolutionarily conserved and may cause tumors if mutated. *Ced-9* encodes a functional homolog of the mammalian proto-oncogene *bcl-2* [Hengartner and Horvitz, 1994].

It is not only the overall growth of bilaterians which is characteristic of the Cambrian explosion, but also differential growth in different parts of the body that was connected with the origin of the new cell types, tissues and organs. Differential growth is also characteristic of tumors, i.e. relative increase in size compared with the other parts of the body. The body wall outgrowths which give rise to limb buds and animal appendages, ectodermal thickenings that give rise to placodes and bags of cells invaginated from the larval epidermis that give rise to imaginal discs in insects may look like and recapitulate these prototype tumors or tumor-like formations in early bilaterians.

10.10 THE NEED FOR A CONTINUOUS SUPPLY OF EXTRA CELL MASSES IN EVOLUTION

The emergence of extra cell masses for evolutionary innovations and morphological novelties did not happen only once in evolutionary history. Set-aside cells may be covergent in situations when larva and adult have different selective regimes [Collins and Valentine, 2001], or in animals with a larger body size [Conway Morris, 1998]. The planktotrophic larvae could have been independently originated at least three times in Bilateria, together with the corresponding set-aside cells. Set-aside cells could be an adaptation associated with metamorphosis [Valentine and Collins, 2000] and a larger body size [Peterson et al., 2005].

In insects, the imaginal discs contain set-aside cells, which are not homologous to the set-aside cells of ciliated feeding larvae of marine animals [Collins and Valentine, 2001].

Beetle horns form from compact discs of epidermal cells that proliferate during the late larval period and evert during the pupal molt [Moczek et al., 2006; Shubin et al., 2009]. Although horn development resembles the development of insect appendage imaginal discs, they do not arise from imaginal discs, but from the novel discs — the new regions of epidermis which at a certain period in evolution began to behave like imaginal discs [Emlen et al., 2007]. For my consideration of the evolutionary role of tumors it is interesting that prepupal horn primordia in different *Onthophagus* species (see e.g. Figure 4 in [Moczek et al., 2006]) look like papillomas [M.A. Zabezhinskiy, personal communication]. Four appendage patterning genes (*exd, hth, dac* and *Dll*) are expressed in prepupal horn primordia [Moczek et al., 2006], suggesting that beetle horns are the product of the co-option and deployment of an appendage patterning program at novel anatomical sites [Shubin et al., 2009] and in papilloma-like outgrowths which provide the extra cell mass for this morphological novelty. The insulin/insulin-like growth factor (IGF) pathway, an ancient pathway which couples the rate of cell proliferation with available nutrients, participates in the development and evolution of beetle horns, and may be responsible for the extreme growth of these exaggerated and sexually selected structures [Emlen et al., 2007, 2012]. This pathway is also involved in tumor processes and therapies targeted at this pathway are being developed [Larsson et al., 2005; Ryan and Goss, 2008].

The neural crest is a key innovation in vertebrate evolution implicated in the development of major phenotypic characters of vertebrates, especially their ectomesenchymal derivatives including bone, cartilage and dentine. Neural crest is the embryonic cell population with two main features: these cells are migratory and multipotent [Donoghue et al., 2008]. Before migration, they undergo an epithelial-to-mesenchymal transition [Baker and Bronner-Fraser, 1997]. These features, common to many tumors, may recapitulate those of prototype tumor-like formations in early vertebrates which might gave rise to neural crests. In humans, there are 47 neural crest–derived cell phenotypes [Vickaryous and Hall, 2006]. Of neural crest–expressed genes, many are vertebrate innovations. Among different tissues, the neural crest exhibits a particularly high rate of expression of novel genes during vertebrate evolution. A remarkable proportion of these genes encode soluble ligands that control neural crest precursor specification into corresponding derivative cell lineages. The expression of an increasing number of novel genes coding for growth factors is not limited to the neural crest: it also occurs in other vertebrate-specific tissues [Martinez-Moralez et al., 2007].

In vertebrates, up to 479 morphological characters could be scored that may be an index of their morphological complexity [Aburomia et al., 2003], and humans have over 400 different specialized cell types [Vickaryous and Hall, 2006]. We may infer that to generate the vertebrate innovations, extra cell masses were needed and supplied at least several hundred times in the course of their evolution. An example of such an innovation which needed extra cell mass is the right ventricle of the vertebrate heart, discussed above in Part 9.2.1. We may further suggest that frequent, ongoing and widespread tumor-like processes provided extra cell masses for the expression of novel genes, the origins of which were also frequent, ongoing and widespread in evolution. The meaningful coincidences of these relatively independent processes at different structural levels might have been "frozen" by natural selection and led to the origin of evolutionary innovations and morphological novelties. Such coincidences should have happened hundreds of times in the evolution of vertebrates. They may still continue in our times, as far as the origin of novel genes and tumor processes are still frequent, ongoing and widespread. If the above reasoning is correct, new technologies to control tumors may be developed on its basis.

10.11 NONADAPTIVE ORIGINS OF ORGANISMAL COMPLEXITY

The increase in complexity of multicellular organisms may be due to nonadaptive processes, and are largely driven by nonadaptive evolutionary forces — at least at the initial stages of complexification. The following arguments are in support of such an assumption. First of all, there is no evidence for the assumption that complexity is caused by adaptive processes. On the contrary, multicellular organisms have reduced population sizes, reduced recombination rates and increased deleterious mutation rates that altogether decrease the efficiency of selection. Because of their relatively small effective number of gene copies per locus at the population level (Ng), multicellular organisms may accumulate genetic changes and expand genomic and biological complexity without selection [Lynch, 2007].

Other authors have also supported the view that the initial rise in body-plan complexity was passive and not connected to external triggers [Valentine, 1994].

This conclusion is in correspondence with the possible evolutionary role of tumors discussed above, because at the initial stages of the genesis of novelties tumors do not play any adaptive role. We may thus overcome the critique based on the gradualism principle that set-aside cells lack selective advantage until they had given rise to adult structures [Wolpert, 1999].

10.12 TUMORS OR COMPLEXITY FIRST?

As it was discussed above in Parts 10.9 and 10.10, cellular proliferative tumor-like processes have been necessary for the origin and evolution of complexity of multicellular organisms since Bilateria. On the other hand, comparative oncology came to the conclusion that tumors are more frequent among the "higher" forms, e.g. insects and vertebrates [Wellings, 1969] and in the evolutionarily more successful groups of organisms, e.g. in teleost fishes compared with cartilaginous fishes [Dawe, 1973] (see discussion in Part 3.1). Armand Leroi and co-authors put forward the arguments that cancer occurs in organs that have undergone recent and pronounced evolutionary change, and represents a by-product of novel adaptation [Leroi et al., 2003; see also Graham, 1992]. I argue that tumor-like processes could participate in such organs' evolutionary origin in the first place, and that tumor growth in more complex organisms recapitulates the original tumor involvement in the evolutionary process. Leroi and co-authors suggest that anti-cancer selection led to a lower incidence of cancer. This was criticized by Frietson Galis and Johan Metz, who presented reasons why cancer risks cannot always be reduced by natural selection [Galis and Metz, 2003].

10.13 TUMORS AS THE SEARCH ENGINE FOR INNOVATIONS AND NOVEL MOLECULAR COMBINATIONS

Pierce called cancer cells the caricature of the normal process of tissue renewal. Neoplastic tissue is a "caricature" of normal tissue in the sense that there is an overproduction of undifferentiated stem cells as compared to differentiated cells. The term caricature means a gross misrepresentation, and proliferation and differentiation are grossly misrepresented in malignant tissues [Pierce et al., 1978; Arechaga, 1993; Pierce, 1993].

In a similar manner, more recent authors call cancer a parody of the normal procession of stem cells to the sequential hierarchy of pluripotent and terminally differentiated blood or tissue cells [Maehaut et al., 2010].

As we discussed earlier, tumors are also considered as atypical or aberrantly developed organs [Hanahan and Weinberg, 2000, 2011; Reya et al., 2001; Brabletz et al., 2005; Boman and Wicha, 2008; Eyler and Rich, 2008; Vermeulen et al., 2008].

Many authors have pointed out that in tumors the abnormal combination of normal cell properties is expressed [Markert, 1968; Coggin and Anderson, 1974]. As we discussed earlier in Part 4.1, many unusual genes and gene sets are activated in tumors.

From the point of view of the hypothesis of evolution by tumor neofunctionalization, all these features suggest that tumors could be regarded as unstable transitory search organs for innovations and expression of evolutionarily novel genes and new combinations of expressed genes which are not possible in established organs and ontogeneses. If tumor-bearing organisms live long enough to leave progeny, advantageous innovations could be supported by natural selection. With the origin of new functions, atypical organs could be stabilized by functional feedbacks, accumulate a larger proportion of differentiated cells and become new organs.

Chapter 11

Experimental Confirmation of Nontrivial Predictions of Evolution by the Tumor Neofunctionalization Hypothesis

The hypothesis of evolution by tumor neofunctionalization makes it possible not only to explain and systematize a vast volume of observational and experimental data but also to formulate nontrivial predictions that can be experimentally tested. I can make two strong predictions: (1) Evolutionarily young and novel genes/sequences are specifically expressed in tumor cells, and (2) Tumors may be selected for new organismal functions.

11.1 EVOLUTIONARILY YOUNG AND NOVEL GENES ARE EXPRESSED IN TUMORS

To study experimentally the prediction concerning the expression of evolutionarily young or novel genes in tumors, two complementary approaches could be used. One is to study the evolutionary novelty of genes/sequences with proven tumor specificity of expression. The other is to study tumor specificity of expression of genes/sequences with proven evolutionary novelty. We performed both approaches and obtained results predicted by the hypothesis.

11.1.1 Analysis of Evolutionary Novelty of Tumor-Specifically Expressed EST Sequences

By the end of the 1990s, a large number of cDNA libraries derived from a vast variety of normal and tumor tissues became available as public databases on the World Wide Web. We decided to carry out a global comparison of cDNA sequences from all tumor-derived libraries with cDNA sequences from all normal tissue-derived libraries. In essence, this was a recapitulation of our previous hybridization experiments with "combined" RNA preparations [Evtushenko et al., 1989; Kozlov et al., 1992], but at a higher methodological level.

The methodology and results of this comparison are published in [Baranova et al., 2001]. The work included the following steps. First, the data of each DNA library were scrutinized to exclude any mistakes in referring to a library as "tumor-derived" or "normal tissue-derived." Using publicly available databases we developed our own database which contained the descriptions of all cDNA libraries classed as "tumorous" or "normal" depending upon cDNA origin. Only rigorously characterized libraries were included, whereas libraries labeled with such terms as "premalignant," "noncancerous pathology," "immortalized cells," or even less definitive descriptions were omitted. Many mistakes and information gaps were found in the source libraries. In total, 2681 libraries were categorized as "tumorous," and 1087 as "normal." The tumorous libraries contained 921,237 expressed sequence tags (EST) in 56,241 clusters (a cluster is an EST set roughly corresponding to a transcription unit), and normal libraries contained 810,097 EST in 53,762 clusters. The number of EST not included in either library type was 500,462 (Figure 11.1).

To find the sequences that are expressed in tumors but not in normal tissues, the normal EST set was subtracted *in silico* from the tumorous EST set. This approach is known as computer-assisted differential display (CDD). A special software tool was developed for our particular purpose [Baranova et al., 2001].

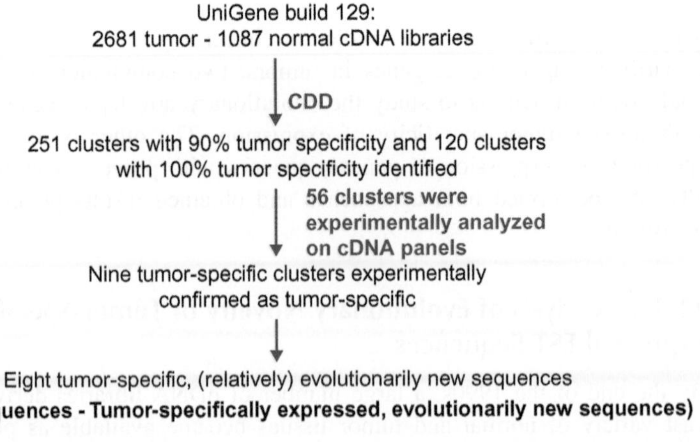

UniGene build 129:
2681 tumor - 1087 normal cDNA libraries

↓ CDD

251 clusters with 90% tumor specificity and 120 clusters
with 100% tumor specificity identified

56 clusters were
experimentally analyzed
on cDNA panels

Nine tumor-specific clusters experimentally
confirmed as tumor-specific

↓

Eight tumor-specific, (relatively) evolutionarily new sequences

(*TSEEN* sequences - Tumor-specifically expressed, evolutionarily new sequences)

FIGURE 11.1   Flow chart describing the discovery of tumor-specifically expressed, evolutionarily new (*TSEEN*) sequences.

The results showed that, in accordance with our prediction, tumors indeed express hundreds of sequences that are not expressed in normal tissues (Figure 11.1). The tumor-specific clusters we found using CDD included several already known tumor markers (e.g. the MAGE family) and no known oncogenes, which is in agreement with protooncogene expression in many normal tissues. About half of the discovered tumor-specific sequences lack long reading frames (i.e. may be referred to noncoding RNA) and defined function [Baranova et al., 2001]. Subsequently, we repeated the procedure of global subtraction of normal libraries from tumorous libraries (global CDD) several times using the modified software tools and expanded sets of tumor libraries (up to 4564) and normal libraries (up to 2304). The results are similar to those discussed above [Galachyants and Kozlov, 2009]. By now, we have at hand hundreds of molecular probes which we can use in studies of the tumor-related and evolutionary specificities of the sequences that we have found.

We carried out *in vitro* experiments intended to confirm that the sequences found with CDD are indeed specifically expressed in tumors. We used cDNA panels from normal and tumor tissues (Clontech Multiple Tissue cDNA (MTC™) panels) and also cDNA obtained at our laboratory. In total, our experiments covered cDNA from tumors of 19 localizations and from 27 normal human tissues including eight human embryo tissues.

The specificity of expression of a defined sequence found with global CDD was checked against the above cDNA panels with PCR using primers specific to the sequence. In total, 56 sequences described in [Baranova et al., 2001] have been studied in this way. Among them, nine were confirmed to be highly tumor-specific [Krukovskaya et al., 2005, 2007; Palena et al., 2007] (Figure 11.1). The sequences that have been confirmed to be tumor-specific are expressed in a vast variety of tumors. For example, the sequence Hs. 202,247 is expressed in 46 tumor samples out of 56 examined and in none of 27 normal tissues. One of the protein products of the sequences that proved to be tumor-specific appeared to be a promising immunogen for antitumor vaccine development [Palena et al., 2007], and the respective US patent has been obtained [Schlom et al., 2012]. However, most of the confirmed tumor-specific sequences appear to be noncoding RNAs.

The nine experimentally confirmed tumor-specific sequences are listed in Table 11.1. These sequences vary in their functional features and annotations (Table 11.2) [Samusik et al., 2011]. Six of the nine sequences are either full-length mRNAs (nos. 1, 3 and 4) or long spliced ESTs (nos. 5, 7 and 8). Three sequences (nos. 2, 6 and 8) are mapped to introns of annotated mRNAs. Interestingly, sequence no. 8 is a three-exon RNA which is mapped into the intron of CACNA2D3 mRNA and is transcribed from the strand opposite to it. Sequences nos. 2 and 6 are short non-spliced ESTs derived from cDNA libraries which had been produced by means of RT with random hexamer primers, so it is impossible to determine *a priori* which strand they were

TABLE 11.1 List of Experimentally Confirmed Tumor-Specific Sequences

No.	Sequence used for Primer Design	UniGene Cluster	Cluster's Anchor Sequence	Cluster Title UniGene
1	NM_005712*	Hs. 285026	AF110315	**HHLA1:** Human HERV LTR Associated 1
2	BE50372S	Hs. 202247	NM_032109	**OTP:** Orthopedia homolog (Drosophila)
3	NM_005987	Hs. 46320	NM_005987	**SPRR1A:** Small Proline-Rich Protein 1A
4	NM_003181	Hs. 389457	NM_003181	**T:** T, brachyury homolog (mouse)
5	AI793334	Hs. 150166	BX119167	Transcribed locus
6	AI792557	Hs. 133107	BC033263	**PVT1:** Pvtl oncogene homolog, MYC activator (mouse)
7	BG822407	Hs. 633957	BX119057	CDNA FU38626 fis, clone HEART2009599
8	DQ445779	Hs. 128594	DQ445779	**CACNA2D3:** Calcium channel, voltage-dependent, alpha 2/delta 3subunit
9	AA166653	Hs. 426704	BC043008	Human ribosomal DNA complete repeating unit

*UniGene cluster identifiers follow UniGene build 210. The clusters' anchor sequence is either RefSeq mRNA or the longest EST of a cluster, which contains most of its exons and represents a probable source transcript.
Source: Samusik et al., 2011, with permission.

transcribed from. Sequence no. 9 is a stand-alone short non-spliced repeat-flanked EST mapped to a spacer region in a highly repetitive rDNA locus.

Sequences nos. 3 and 4 encode SPRRA1 and T proteins, respectively. Sequence no. 3 was shown to be overexpressed in many epithelial tumors [Sark et al., 1998].

These nine experimentally confirmed tumor-specific sequences were studied for their evolutionary novelty using molecular-biological techniques, comparative genomics analysis, the search for orthologous sequences and sequence conservation analysis [Kozlov et al., 2006; Samusik et al., 2009, 2011].

Analysis of orthologs of tumor-specifically expressed sequences shows that the majority of the studied sequences are relatively evolutionarily new.

TABLE 11.2 Annotations of Analyzed Sequences

	Protein-coding mRNA	Long Spliced Transcripts	Short Unspliced Transcripts
Stand-alone transcripts	**No. 1** (HHLA1, Hs. 285026) **No. 3** (SPRR 1A, Hs. 46320) **No. 4** (T, Brachvurv, Hs. 389457)	**No. 7** (TrSeq, Hs. 633957) **No. 5** (TrSeq, Hs. I50166)	**No. 9** (rRNA repeating unit, Hs. 426704)
Mapped to introns of other mRNA		**No. 8** (CACNA2D3, Hs. 128594)*	**No. 2** (OTP, Hs. 202247) **No.6** (PVTI, Hs. 133107)

*The sequence is transcribed from the genomic region corresponding to intron of CACNA2D3. The coding strand is opposite to CACNA2D3.
Source: Samusik et al., 2011, with permission.

All the orthologs of sequence no. 8, except chimp, lack the sequence that corresponds to two or more exons in human mRNA. The larger part of the chimp sequence orthologous to human exons 1 and 2 is inverted with respect to the human sequence. The region specific to human and chimp genomes is a result of insertion of a 5 kb sequence, homologous to HERV-H and containing viral LTRs.

We observed a similar pattern in the orthologs of sequence no. 1 that corresponds to the human HHLA1 gene. None of the inspected genomes, except for chimp, contains the homologous sequence to first exon of HHLA1. The larger part of the "novel" exon that is specific to chimps and humans consists of an HERV-H sequence containing an LTR, and is a result of an insertion of a partial endoretrovirus sequence into the given genomic locus in the common ancestor of humans and chimps.

The open reading frame (ORF) of HHLA is present only in humans and chimps, and the ORF of intronic transcript CACNA2D3 is present in primates but not in other mammalians.

A BLAST search for homologs of sequence no. 9 revealed them only in the chimp genome as well as in humans. In this case we couldn't identify the orthologous sequence as the ribosomal genes have multiple copies in the human and chimp genomes. Nevertheless we can claim that sequence no. 9 is specific to higher primates.

We found orthologs of sequences nos. 2, 3, 5 and 6 in the mammalian genomes only. Orthologs of sequence no. 7 were found in the genomes of mammals and chickens. However, the chicken otholog sequence is ~300 bp long and has a weak homology with human sequence no. 7. In fish genomes we found orthologs only of sequence no. 4.

We estimated the substitution rate per site per year for pair-wise alignments between human exon sequences and identified corresponding orthologs. We used the two-parameter Kimura model [Kimura, 1980] to estimate the substitution rate per site (K) and its variance. We compared the rates obtained for the sequences of interest to the rates in several groups of annotated genomic sequences or sites. Ka/Ks ratios were also estimated with existing methodology [Comeron, 1995, 1999]. We studied six sequences for which the orthologs had been identified. It is established that the conservation of the nucleotide sequences is a consequence of negative selection and indicates the presence of the function.

Noncoding non-spliced sequences nos. 2 and 6 that are mapped to introns of other mRNA both appear to evolve at nearly neutral rates, as follows from comparison to dog, rat and mouse (respective rates in these sequences are close to substitution rates of "4-d sites" and "ancestral repeats"). A similar situation is observed in case of the long spliced sequence no. 5. The substitution rates compared to orthologs in dog, rat and mouse is not different from the neutral control. Excluding rodents, the substitution rate in sequence no. 7 is higher than the substitution rates in RefSeq exons and is similar to the neutral rates. We found the highly conserved 40-bp region in the first exon of sequence no. 7. Interestingly, this sequence is the only one among nine studied tumor-specifically expressed sequences to have an orthologous region in chickens. The chicken otholog sequence is $\sim$300 bp long and has a weak homology with human sequence no. 7. The chicken ortholog is mapped to the human genome region adjacent to but non-overlapping with the highly conserved region. However, the rest of sequence no. 7 ($\sim$600 bp) appears to have evolved neutrally.

The frequencies of nucleotide substitutions in sequences nos. 3 and 4 coincide with those in RefSeq exons. Mean Ka/Ks between human and chimp, rhesus, rat, dog and mouse in those sequences was 0.13 ± 0.02 and 0.12 ± 0.02, respectively, with significant difference from 1.0 for each organism. Hence the exon regions of protein-coding sequences nos. 3 and 4 are conserved.

Sequence no. 3 corresponds to the *SPRRA1* gene. Unlike sequence no. 4, whose orthologs are present in all analyzed non-human genomes, the orthologs of sequence no. 3 are found only in mammals. That is why sequence no. 3 may be considered as relatively evolutionarily novel. This sequence has already acquired a function and plays a role in the maturation of cells in cornfield epithelia [Sark et al., 1998].

So we were able to show that eight of the nine tumor-specifically expressed sequences are either evolutionarily new (three of them are present only in primates) or relatively young and evolve neutrally [Kozlov et al., 2003, 2006; Samusik et al., 2007, 2009, 2011]. I suggest calling such sequences tumor-specifically expressed, evolutionarily new sequences, or *TSEEN* sequences (Figure 11.1).

Sequences nos. 1 and 8 both contain ORFs, although the corresponding proteins haven't been shown experimentally. These sequences are similar to recently described human *de novo* protein-coding genes [Knowles and McLysaght, 2009]. Three novel human protein-coding genes have been shown by these authors to originate from noncoding DNA since the divergence with chimp. These genes have no protein-coding homologs in any other genome. The existence of protein-coding genes is supported by expression and proteomic data. As far as corresponding proteins have not been shown for sequences nos. 1 and 8, these sequences may represent the earlier stage of the novel gene origin comparing to those described by D.G. Knowles and A. McLysaght. These and other sequences described in this section (besides protein-coding sequences nos. 3 and 4 with established functions) may represent protogenes at different stages of their evolution towards novel genes with protein- or RNA-related functions. Sequence no. 7 can illustrate such a transition.

Before we initiated our study, little was known about the noncoding sequence no. 7 (Hs.633957). Neither the literature nor automated annotations provided any insights about the function of the highly conserved 40-bp region of its first exon. We speculated that an unknown function that is realized at the RNA level can be connected with the conservative region of sequence Hs.633957.

11.1.2 *ELFN1-AS1* — a Novel Primate Gene with Possible MicroRNA Function Expressed Predominantly in Tumors

The human transcribed locus *LOC100505644 uncharacterized LOC100505644* resides in the 7th chromosome and corresponds to the UniGene EST cluster Hs.633957. It was found by our group to be expressed in a tumor-specific manner by *in silico* analysis [Baranova et al., 2001]. Later we supported these data experimentally: specific transcripts of the locus *LOC100505644 uncharacterized LOC100505644* were detected in tumors of various histological origin, but not in most of the healthy tissues [Krukovskaya et al., 2005; Polev et al., 2009, 2011a].

As far as nothing was known about the functionality of the locus *LOC100505644 uncharacterized LOC100505644*, we performed a comprehensive study of it using a combined computational and experimental approach [Polev et al., 2011b, 2013]. We obtained experimental and *in silico* evidence that locus *LOC100505644 uncharacterized LOC100505644* is a stand-alone gene which has its own promoter and capability for alternative splicing. However, only one splicing isoform is predominant. The gene was assigned the name *ELFN1-AS1* (see below for the explanation of the gene name). We found that the TSS fell into a chromatin region rich in promoter- and enhancer-specific histone modifications. A GC-box, which is a common core promoter element, was identified right before the TSS. It coincides with the DNaseI hypersensitivity site, suggesting that a core promoter responsive for Sp1 TFs is there. About 20 transcription factors were found to be specifically associated with the promoter region of the gene in different cells. Several of the DNA-binding TFs have putative binding sites within the region. These data suggest that the activity of the promoter identified in the present work is required for *ELFN1-AS1* expression. We found eight E-boxes in the promoter proximal region and all of them rest in chromatin regions with which c-Myc and Max are associated in specific cells. These data indicate that *ELFN1-AS1* is a c-Myc-responsive gene.

The analysis of the secondary structure of the major transcript variant revealed a hairpin-like structure characteristic of precursor microRNAs. A 68-nucleotide hairpin was identified (Figure 11.2A). It is formed by nucleotides 3502−3569 of the putative gene and resembles the hairpins of known pre-miRNAs. The 5'-strand of the hairpin stem holds a 22 nt fragment which is complementary to a region of DPYS (dihydropyrimidinase) mRNA (Figure 11.2B). This 22 nt site is located in the coding region of the mRNA in close proximity to the 3' UTR. These data support the possibility that *ELFN1-AS1* may code for a miRNA. This is further supported by the incomplete complementarity of the putative miRNA and its target site with unpaired nucleotides in position 9, and the presence of adenine in the first position of the target site. Such features are favorable for miRNAs' action as suppressors at the RNA level [Lewis et al., 2005; Selbach et al., 2008]. Seed-sites of the miRNAs may serve as initiators for the miRNA-target

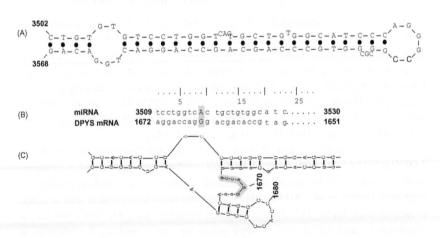

FIGURE 11.2 A hairpin-like secondary structure of ELFN1-AS1 miRNA (A); the miRNA 22 nt fragment complementary to a region of DPYS mRNA (B); putative miRNA and site counterpart in the DPYS mRNA (C).

interaction through fast target recognition [Rajewsky, 2006]. In compliance with this, the putative miRNA seed site counterpart in the DPYS mRNA is single-stranded while it is surrounded with double-stranded RNA regions (Figure 11.2C). DPYS participates in pyrimidine catabolism [van Kuilenburg et al., 2010] and is active in the liver and in solid tumors [Naguib et al., 1985]. According to our data, the liver is one of the few normal tissues where weak expression of the gene is observed [Krukovskaya et al., 2005; Polev et al., 2009, 2011a]. Finally, we found that our putative miRNA appeared in the course of primate evolution at about the same time as did its putative target site in DPYS mRNA, though the *DPYS* gene itself is conservative in vertebrates. Thus all data point to the miRNA function of *ELFN1-AS1* with DPYS mRNA being its primary target (Figure 11.2).

The origin of the gene is of particular interest. According to our data it arose *de novo* from an intronic region of a conservative gene *ELFN1* (NCBI Ref. Seq. NM_001128636.2) in primate lineage. Non-primate mammals have only small regions of homology with human *ELFN1-AS1*. Homologous sequences of this gene were identified by us in all primates, but the DNA sequence from the representative of suborder Strepsirrhini *Otolemur garnettii* has more than 50% differences from its human counterpart and forms an outgroup on the phylogenetic tree. It is also different from other primates in that it has no GC-box, GABP- and E2F-binding sites in the region corresponding to human TSS. Thus *ELFN1-AS1* could become transcriptionally active after divergence of Strepsirrhini and Haplorhini primates. It is noteworthy that all the Haplorhini primates have a region with five or more E-boxes downstream of the the the DS site. This suggests that the *ELFN1-AS1* gene since its origin could be c-Myc-responsive.

Taken together, our data indicate that human transcribed locus *LOC100505644 uncharacterized LOC100505644* contains a gene for some noncoding RNA, likely a microRNA. It was assigned a gene symbol *ELFN1-AS1*, ELFN1 antisense RNA 1 (non−protein-coding), which was approved by the Human Gene Nomenclature Committee. This gene combines features of predominant expression in tumors and evolutionary novelty [Polev et al., 2011b, 2014]. I suggest calling such genes tumor-specifically expressed, evolutionarily new genes, or *TSEEN* genes (Figure 11.3).

11.1.3 *PBOV1*: A Human Gene of Recent *De Novo* Origin with a Highly Tumor-Specific Expression Profile

In the study of *PBOV1* gene we used the approach complementary to that described in Parts 11.1.1 and 11.1.2, i.e. we studied the tumor specificity of a gene that was supposedly evolutionarily new [Krukovskaya et al., 2010; Samusik et al., 2013].

We first noticed this gene, as mentioned, among 12 human genes without orthologs in the mouse and dog genomes in the paper of Clamp and

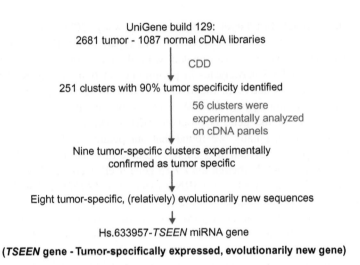

FIGURE 11.3   Flow chart describing the discovery of tumor-specifically expressed, evolutionarily new (*TSEEN*) putative miRNA gene.

co-authors [Clamp et al., 2007]. *PBOV1* (*UROC28, UC28*) is a human protein-coding gene with a 2501 bp single-exon mRNA and 135aa ORF. The gene was originally characterized by An and co-workers [An et al., 2000].

First of all, we studied the evolutionary novelty of this gene more carefully. We performed a detailed comparison of the protein-coding sequence of *PBOV1* with 34 genomes of placental mammals. For each genome, we computed the fraction of the human coding sequence that can be aligned with it. Based on the presence of frame-shift mutations and stop-codons, we deduced the fraction of the human protein sequence that was homological to the putative protein obtained by translation of the target sequence in the other species. We mapped the results to the mammalian evolutionary tree and indicated the key evolutionary steps that led to the appearance of human *PBOV1* (Figure 11.4).

We found that the coding sequence of *PBOV1* is poorly conserved in mammalian evolution and originated *de novo* in primate evolution through a series of frame-shift and stop codon mutations. Consequently, 80% of the protein sequence is unique to humans. We have tested the *Ka/Ks* ratio both in pairwise alignments and in a multiple alignment of all primate sequences syntenic to the human coding sequence and didn't find any significant differences from 1.0, indicating that the amino acid sequence evolved neutrally. PBOV1 protein lacks any annotated or predicted domains and over 60% of its sequence is predicted to be disordered. These findings altogether strongly suggest that human PBOV1 is a protein of a very recent *de novo* evolutionary origin (Figure 11.4).

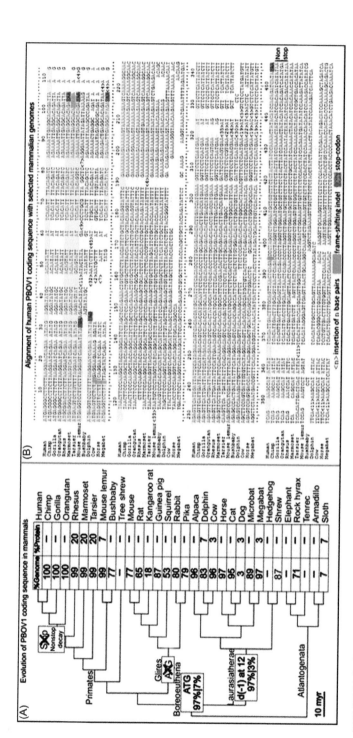

FIGURE 11.4 Evolution of PBOV1 protein-coding sequence in mammals. *From open access paper [Samusik et al., 2013]. Copyright of authors.*

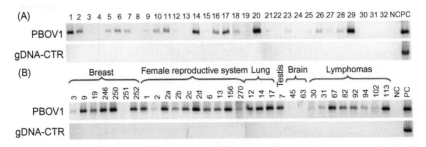

FIGURE 11.5 PBOV1 expression measured by PCR in cDNA panels from human tumors. DNA contamination was controlled using gDNA-CTR primers targeting an exon-intron junction of HERC1 gene. (A) Tumor cDNA Panel (BioChain Institute, USA): 1—Brain medulloblastoma, with glioma, 2—Lung squamous cell carcinoma, 3—Kidney granular cell carcinoma, 4—Kidney clear cell carcinoma, 5—Liver cholangiocellular carcinoma, 6—Hepatocellular carcinoma, 7—Gallbladder adenocarcinoma, 8—Esophagus squamous cell carcinoma, 9—Stomach signet ring cell carcinoma, 10—Small intestine adenocarcinoma, 11—Colon papillary adenocarcinoma, 12—Rectum adenocarcinoma, 13—Breast fibroadenoma, 14—Ovary serous cystoadenocarcinoma, 15—Fallopian tube medullary carcinoma, 16—Uterus adenocarcinoma, 17—Ureter papillary transitional cell carcinoma, 18—Bladder transitional cell carcinoma, 19—Testis seminoma, 20—Prostate adenocarcinoma, 21—Malignant melanoma, 22—Skeletal muscle malignancy fibrous histocytoma, 23—Adrenal pheochromocytoma, 24—Non-Hodgkin's lymphoma, 25—Thyroid papillary adenocarcinoma, 26—Parotid mixed tumor, 27—Pancreas adenocarcinoma, 28—Thymus seminoma, 29—Spleen serous adenocarcinoma, 30—Hodgkin's lymphoma, 31—T cell Hodgkin's lymphoma, 32—Malignant lymphoma. NC—PCR with no template, PC—PCR with human DNA. (B) PBOV1 expression in clinical tumor samples. PBOV1 is expressed in breast cancer (9–250), ovary cancer (1, 6), cervical cancer (2, 13), endometrial cancer (156, 270), lung cancer (12, 14, 17), seminoma (7), meningioma (63), non-Hodgkin's lymphomas (67, 82, 92, 102, 113). *From open access paper [Samusik et al., 2013]. Copyright of authors.*

After establishing the evolutionary novelty of the *PBOV1* gene, we studied the specificity of its expression in tumors and normal tissues. *PBOV1* has been previously reported to be overexpressed in prostate, breast and bladder cancers [An et al., 2000]. We studied the expression of *PBOV1* using PCR on panels of cDNA from various normal and tumor tissues. We profiled *PBOV1* expression in multiple cancer and normal tissue samples and found that the gene had a highly tumor-specific expression profile. It was expressed in 20 out of 34 tumors of various origins (Figure 11.5) but was not expressed in any of the normal adult or fetal human tissues that we tested (Figure 11.6). The most outstanding feature of this result is that tumor specificity of *PBOV1* expression was predicted by its evolutionary novelty.

We found that unlike cancer/testis antigen genes *PBOV1* was expressed from a GC-poor TATA-containing promoter which was not influenced by DNA methylation and is not active in the testis. We found evidence that *PBOV1* activation in tumors depended on sex hormone receptors, C/EBP transcription factors and the Hedgehog signaling pathway. Although the

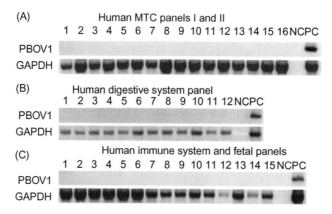

FIGURE 11.6 Expression of PBOV1 and GAPDH (positive control) measured by PCR in cDNA panels from human normal tissues. (A) Human MTC Panel I (1–8), Human MTC Panel II (9–16): 1—brain, 2—heart, 3—kidney, 4—liver, 5—lung, 6—pancreas, 7—placenta, 8—skeletal muscle, 9—colon, 10—ovary, 11—peripheral blood leukocyte, 12—prostate, 13—small intestine, 14—spleen, 15—testis, 16—thymus; (B) Human Digestive System MTC Panel: 1—cecum, 2—colon, ascending 3—colon, descending 4—colon, transverse 5—duodenum, 6—esophagus, 7—ileocecum, 8—ileum, 9—jejunum, 10—liver, 11—rectum, 12—stomach; (C) Human Immune System MTC Panel (1–7), Human Fetal MTC Panel (8–15): 1—bone marrow, 2—fetal liver, 3—lymph node, 4—peripheral blood leukocyte, 5—spleen, 6—thymus, 7—tonsil, 8—fetal brain, 9—fetal heart, 10—fetal kidney, 11—fetal liver, 12—fetal lung, 13—fetal skeletal muscle, 14—fetal spleen, 15—fetal thymus; A-C: NC—PCR with no template, PC—PCR with human DNA. *From open access paper [Samusik et al., 2013]. Copyright of authors.*

PBOV1 protein has recently originated *de novo* and thus has no identifiable structural or functional signatures, a missense SNP in it has been previously associated with an increased risk of breast cancer. Using publicly available data we found that a higher level of *PBOV1* expression in breast cancer and glioma samples was significantly associated with a positive disease outcome. We also found that *PBOV1* was highly expressed in primary but not recurrent high-grade gliomas, suggesting that immunoediting against *PBOV1*-expressing cancer cells might occur over the course of disease. We propose that *PBOV1* is a novel tumor suppressor which might act by provoking a cytotoxic immune response against cancer cells that express it. We speculate that this property might be a source of phenotypic feedback that facilitated *PBOV1* fixation in human evolution [Samusik et al., 2013].

11.1.4 The Evolutionary Novelty of Human Cancer/Testis Antigen Genes

In order to be inherited in progeny generations, novel genes should originate in germ cells. Indeed, all the data discussed in Chapter 8 prove that the processes of the origin of novel genes in germ cells are actively going on because novel genes are inherited in progeny generations. Some peculiarities of germ cells' physiology support the process of the origin of new genes, e.g. promiscuity of gene expression in spermatogenic cells [Schmidt, 1996; Kleene, 2005]. Novel genes originated through different mechanisms (retrogenes, segmental duplicates, chimeric and *de novo*—emerged genes) are all expressed in the testis [Betran et al., 2002; Paulding et al., 2003; Levine et al., 2006; She et al., 2008; Heinen et al., 2009; Kaessmann et al., 2009; reviewed in Kaessmann, 2010]. This led to the suggestion that the testis may play a special "catalyst" role in the birth and evolution of new genes [Kaessmann, 2010].

On the other hand, a new class of genes — cancer/testis or cancer/germline antigen genes — has been recently discovered, the major characteristic of which is expression predominantly in the testis and a variety of tumors. My prediction, based on the hypothesis of evolution by tumor neofunctionalization, was that cancer/testis antigen genes are evolutionarily young or novel. This prediction was studied in our paper [Dobrynin et al., 2013], the major points of which are presented below.

11.1.4.1 The Class of Cancer/Testis Antigen Genes

Cancer/testis antigen genes (CTA or CT genes) code for a subgroup of tumor antigens expressed predominantly in the testis and different tumors. CT antigens may be also expressed in the placenta and in female germ cells [Zendman et al., 2003a,b; Simpson et al., 2005; Chen et al., 2006]. Surprisingly, some CT antigens are also expressed in the brain [Hofmann et al., 2008].

The methods used for the discovery of CT genes include immunological screening methods [van der Bruggen et al., 1991], SEREX technology (serological identification of antigens by recombinant expression cloning) [Sahin et al., 1995], expression database analysis [Scanlan et al., 2002, 2004], massively parallel signature sequencing [Chen et al., 2005a] and other approaches. The fact that many CT antigens have been identified using SEREX technology suggests that they are highly antigenic [Sahin et al., 1995; Cheng et al., 2011].

The first CT gene to be discovered was the *MAGEA1* gene coding for an antigen of human melanoma [van der Bruggen et al., 1991]. This gene belongs to a family of 12 closely related genes clustered at Xq28. A second cluster of *MAGE* genes, *MAGEB*, was discovered at Xp21.3, and the third, encoding *MAGEC* genes, is located at Xq26—27. The expression of *MAGEA* — *MAGEC* genes (*MAGE-I* subfamily) is restricted to testis and cancer, whereas more distantly related clusters *MAGED* — *MAGEL* (subfamily *MAGE-II*) are expressed in many normal tissues. *MAGE-I* genes are of relatively recent origin, and *MAGE-II* genes are relatively more ancient.

For example, *MAGE-D* genes are conserved between man and mouse. One of these genes corresponds to the founder member of the family, and the other *MAGE* genes are retrogenes derived from the common ancestral gene [Chomez et al., 2001; Simpson et al., 2005; Katsura and Satta, 2011].

To date, the CTDatabase (http://www.cta.lncc.br) includes 265 CT genes and 149 CT gene families. More than half of them are located on the X chromosome (CT-X genes) [Hofmann et al., 2008]. The analysis of the DNA sequence of the human X chromosome predicts that approximately 10% of the genes on the X chromosome are of the CT antigen type [Ross et al., 2005]. Non-X CT genes are distributed throughout the genome and are represented mainly by single-copy genes [Simpson et al., 2005; Cheng et al., 2011; Fratta et al., 2011].

In a normal testis, CT-X genes are expressed in proliferating germ cells (spermatogonia). Non-X CT genes are expressed during later stages of germ-cell differentiation, i.e. spermatocytes [Simpson et al., 2005]. Among human tumors, CT antigens are expressed in melanoma, bladder cancer, lung cancer, breast cancer, prostate cancer, sarcoma, ovarian cancer, hepatocellular carcinoma, hematologic malignancies, etc. [Hofmann et al., 2008; Caballero and Chen, 2009; Cheng et al., 2011; Fratta et al., 2011]. Genome-wide analysis of 153 cancer/testis genes' expression has led to their classification into testis-restricted (39), testis/brain-restricted (14) and testis-selective (85) groups of genes, the latter group showing some expression in somatic tissues. The majority of testis-restricted genes are CT-X genes (35 of total 39 testis-restricted groups). Non-X CT genes are more broadly expressed [Hofmann et al., 2008].

Multiple CT antigens are often co-expressed in tumors, suggesting that their gene expression program is coordinated [Sahin et al., 1995, 1998; Simpson et al., 2005]. CT gene expression is controlled by epigenetic mechanisms which include DNA methylation and histone post-translational modifications [Fratta et al., 2011]. Other mechanisms of CT gene regulation include sequence-specific transcription factors and signal transduction pathways such as activated tyrosine kinases [Akers et al., 2010].

The functions of CT-X genes are largely unknown. On the contrary, more is known about functions of non-X CT genes which are associated with meiosis, gametogenesis and fertilization. Non-X CTs are conserved during evolution [Hofmann et al., 2008; Caballero and Chen, 2009; Cheng et al., 2011; Fratta et al., 2011].

CT-X genes tend to form recently expanded gene families, many with nearly identical gene copies [Zendman et al., 2003a,b; Kouprina et al., 2004b; Chen et al., 2005a; Simpson et al., 2005; Chen et al., 2006; Caballero and Chen, 2009].

Rapid evolution of cancer/testis genes on the X chromosome has been demonstrated. Comparison of human:chimp orthologs of these genes has shown that they diverge faster and undergo stronger positive selection than those on the autosomes or than control genes on either the X chromosome or autosomes [Stevenson et al., 2007].

The prevalence of large, highly homologous inverted repeats (IRs) containing testes genes on the X- and Y-chromosomes was described in humans and great apes [Skaletsky et al., 2003; Warburton et al., 2004]. CT-X gene families are also located in direct or inverted repeats [Chen et al., 2006].

The study of clusters of homologous genes originated by gene duplication roughly after the divergence of the human and rodent lineages discovered several families of CT genes among recent duplicates [IHGSC, 2004].

In the other paper, the authors also studied recent duplications in the human genome, and found that CT genes were represented in this gene set, including the family of *PRAME* (a preferentially expressed antigen of melanoma) genes. *PRAME* genes are located on chromosome 1 and expressed in the testis and in a large number of tumors [Birtle et al., 2005]. Duplicated *PRAME* genes are hominin-specific, having arisen in the human genome since the divergence from chimps. The *PRAME* gene family also expanded in other Eutheria. Chimps and mice have orthologous *PRAME* gene clusters on their chromosomes 1 and 4, respectively [Birtle et al., 2005; Chang et al., 2011]. The *PRAME* gene family has experienced considerable positive selection in primates [Demuth and Hahn, 2009].

The *SPANX-A/D* gene subfamily of cancer/testis-specific antigens evolved in the common ancestor of the hominoid lineage after its separation from orangutan. The coding sequences of the *SPANX* genes evolved rapidly, faster than the intron and the 5' untranslated region. Evolution rates of both synonymous and nonsynonymous codon positions were accelerated. The mechanism of expansion of *SPANX* genes was segmental DNA duplications. The authors conclude that positive selection in primate lineage was involved in the evolution of the *SPANX* gene family. But *SPANX* genes are expressed in cancer cells and highly metastatic cell lines from melanomas, bladder carcinomas and myelomas [Kouprina et al., 2004b], which raises the question of the reasons for their positive selection. Earlier Southern blot and database analyses have detected *SPANX* sequences only in primates [Zendman et al., 2003a]. *SPANX-N* is the ancestral form, from which the *SPANX-A/D* subfamily evolved in the common ancestor of hominoids approximately 7 million years ago [Kouprina et al., 2004b, 2007].

The *GAGE* cancer/testis antigen gene family contains at least 16 genes which are encoded by an equal number of tandem repeats. All *GAGE* genes are located at Xp11.23. *GAGE* genes are highly identical and evolved under positive selection, which supports their recent origin [Gjerstorff and Ditzel, 2008; Liu et al., 2008].

The *XAGE* family of cancer/testis antigen genes belongs to the superfamily of GAGE-like CT genes. It is located on chromosome Xp11.21 – Xp11.22. Three XAGE genes are described, as well as several splice variants of *XAGE-1* [Zendman et al., 2002; Sato et al., 2007].

The *CT45* gene family was discovered by massively parallel signature sequencing. It includes six highly similar (>98%) genes that are clustered in tandem on chromosome Xq26.3. The *CT45* antigen is expressed in Hodgkin's lymphoma and in other human tumors [Chen et al., 2005a, 2009, 2010a; Heidebrecht et al., 2006].

The *CT47* cancer/testis multicopy gene family is located on chromosome Xq24. Among normal tissues, it is expressed in the testis and (weakly) in the placenta and brain. In tumors, its expression was found in lung cancer and esophageal cancer. The *CT47* family members are characterized by high (>98%) sequence homology. The chimp is the only other species in which a gene homologous to *CT47* was found [Chen et al., 2006].

The ectopic expression of germ cell genes in somatic tumor cells was also recently described in *Drosophila* brain cancer [Janic et al., 2010]. Malignant brain tumor signature (MBTS) in *Drosophila* includes overexpression of 102 genes. Transcription profiles of other *Drosophila* neoplasms are different from MBTS that can be used to distinguish distinct tumors. Inactivation of some of these germline genes suppressed malignant growth. Normally, these genes are required for germline stemness, fitness or longevity. Orthologs of some of these genes were also expressed in human somatic tumors.

11.1.4.2 The Results of in silico Study of the Evolutionary Novelty of Cancer/Testis Genes

The hypothesis of the expression of evolutionarily novel genes in tumors explains this otherwise strange cancer–testis association paradox: as far as the origin of evolutionarily novel genes is connected with their expression in testis, cancer/testis genes are novel genes which are expressed both in testis and in tumors.

So my prediction was that cancer/testis antigen genes should be evolutionarily new or young genes. In order to prove this prediction, we studied the presence of genes orthologous to human cancer-testis genes in human lineage [Dobrynin et al., 2013]. We performed this analysis separately for genes located on the X chromosome and autosomal cancer/testis genes, as far as extensive traffic of novel genes has been described for the mammalian X chromosome [Betran et al., 2002; Emerson et al., 2004; Levine et al., 2006].

To assess the evolutionary novelty of CT genes we applied the NCBI's HomoloGene tool. To construct the clusters of homologs, the HomoloGene program uses information from blastp, phylogenetic analyses and syntheny information when it is possible. Cutoffs on bits per position and *Ks* values are set to prevent unlikely "orthologs" from being grouped together. These cutoffs are calculated based on the respective score distribution for the given groups of organisms [Sayers et al., 2012].

We searched for homologs of each of the CT genes among annotated genes in several completely sequenced eukaryotic genomes and built distributions of all CT-X genes, all autosomal CT genes, all human CT genes and all annotated protein-coding genes from the human genome according to the origin of their homologs in 11 taxa of human lineage. We have shown that 31.4% of CT-X genes are exclusive to humans and 39.1% of CT-X genes have homologs originating in Catarrhini and Homininae. Thereby the majority of human CT-X genes (70.5%) are novel or young for humans (Figure 11.7). Our data are in good correspondence with evidence obtained

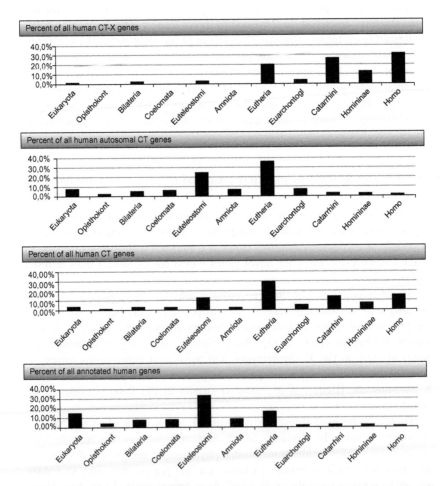

FIGURE 11.7 The proportion of CT-X genes, autosomal CT genes, all human CT genes and all annotated human genes with homologs originated in difference taxa of *H. sapiens* lineage. *Source: open access paper [Dobrynin et al., 2013]. Copyright of authors.*

by other groups on the rapid expansion of certain CT-X gene families and the high homology of their members, which suggest their recent origin.

Altogether 36.7% of all human CT genes originated in Catarrhini, Homininae and humans. We have also found that 30% of all human CT genes originated in Eutheria (Figure 11.7). These CT genes acquired functions in Eutheria. This indicates the importance of processes in which tumors and CT antigens were involved during the evolution of Eutheria (see, for example, the discussion of the eutherian placenta in Parts 5.6, 7.5 and 12.3). CT genes originating in Eutheria are located mainly on autosomes. CT genes originating in Catarrhini, Homininae and humans are located predominantly on the X chromosome. This difference is probably related to important events in the evolution of the mammalian X chromosome since the origin of Eutheria [Lahn and Page, 1999], especially the acquisition of a special role in the origin of novel genes [Kaessmann, 2010].

Thus the majority of CT-X genes are either novel or young for humans, and the majority of all human CT genes (>70%) originated during or after the origin of Eutheria (Figure 11.7). These results, which are statistically significant, suggest that the whole class of human CT genes is relatively evolutionarily new [Dobrynin et al., 2013]. Although the data on the evolutionary novelty of some groups of CT genes already existed in the literature, as discussed above, our study [Dobrynin et al., 2013] was the first systematic study of the evolutionary novelty of all CT genes to show that the whole class of CT genes is relatively evolutionarily new.

The conclusion about the relative evolutionary novelty of the whole class of CT genes confirms the prediction of evolution by tumor neofunctionalization hypothesis about the expression of evolutionarily young and novel genes in tumors. The expression of cancer/testis genes in tumors may be considered a natural phenomenon, not an aberrant process as was suggested by many authors (e.g. [Simpson et al., 2005; Caballero and Chen, 2009; Akers et al., 2010; Chang et al., 2011; Cheng et al., 2011]).

11.1.5 The Phenomenon of Tumor-Specifically Expressed, Evolutionarily Novel (*TSEEN*) or *carcino-evo-devo* Genes

In our studies of the evolutionary novelty of tumor-specifically expressed genes we used two complementary approaches: we studied the evolutionary novelty of genes with established tumor specificity of expression, and we studied tumor specificity of expression of genes with established evolutionary novelty. We studied single locus genes and the complex class of CT genes with many gene families. Using these different approaches, we have been able to describe many genes with tumor-specific expression which are also evolutionarily novel or young, or evolutionarily young and novel genes which are tumor-specifically expressed. Therefore I suggest considering the expression of evolutionarily young or novel genes in tumors as a new biological phenomenon, a phenomenon of *TSEEN* (tumor-specifically expressed, evolutionarily novel) genes. It is similar to the phenomenon of carcinoembryonic antigens. The important difference from carcinoembryonic antigens, though, is that *TSEEN* genes were predicted by the hypothesis of evolution by tumor neofunctionalization. By analogy with the term "carcinoembryonic" and in accordance with the evolution by tumor neofunctionalization hypothesis, these genes may also be called "*carcino-evo-devo*" genes.

We have also described tumor-specifically expressed, evolutionarily new (*TSEEN*) sequences (Figure 11.1) which look like proto-genes, i.e. gene precursors which have not yet acquired functions and evolve neutrally. Expression of proto-genes, novel and young genes in tumors may represent different stages of the origin of a new gene and novel organismal functions (which are not related to tumor progression) in multicellular organisms.

11.2 ARTIFICIALLY SELECTED TUMORS

I was interested in the nature of headgrowths, or hoods, of goldfishes. Based on the hypothesis of evolution by tumor neofunctionalization I inferred that these outgrowths could be benign tumors which were selected for during hundreds of years of breeding. This supposition was studied in a special article [Kozlov et al., 2012]. The major arguments of this work are presented below.

11.2.1 The Goldfish Hood is a Benign Tumor

Comparative oncology studies tumors and tumor-like processes in different species of organisms. These studies are important for understanding the general biologic aspects of cancerogenesis, including the role of exogenous and endogenous factors in tumor development. Comparative oncology was discussed in more detail in Chapter 3. Fishes represent one of the most promising groups for comparative oncological studies. Fishes are the largest group of vertebrates which includes more than 20,000 species [Groff, 2004]. They are used for studies of spontaneous and induced cancerogenesis [Wellings, 1969; Groff, 2004], tumor development as related to environmental contamination [Grizzle et al., 1984; Lindesjoo and Thulin, 1994] and the role of genetic factors in tumor development [Gordon, 1927; Kosswing, 1928; Ahuja and Anders, 1976; Anders et al., 1984; Anders, 1991].

Aquarium fishes represent a convenient model for studies of different aspects of fish neoplasia. A history of more than 1500 years of breeding *Carassius auratus* in China and Japan has resulted in many varieties of goldfishes. Some of these varieties [Oranda and Red cap Oranda, Lionhead, Ranchu] develop headgrowths, or hoods, which sometimes are pigmented. These hoods may grow so large that they may cover the whole surface of the head and make it hard for the fishes to breathe and feed. The biological nature of these outgrowths was not elucidated.

We studied the morphology and dynamics of hood growth in Oranda and Lionhead goldfishes in order to understand the nature of these headgrowths [Kozlov et al., 2012]. We performed macroscopic and microscopic studies of adult hoods, as well as the studies of the dynamics of the hood growth. For the latter purpose, 100 baby fishes obtained by hybridization of Oranda and Fantail goldfishes were studied. The fish were under permanent observation during the period of more than two years.

During the follow-up, individual fishes were periodically taken for histo-logical study of the head skin. Fishes with frank hoods, feebly marked hoods and without hoods could be identified after macroscopic examination (Figure 11.8A−D). The hoods were located in the occipital region of the head. They were often pigmented and rose over the surface of the skin. The surface of the hoods was uneven, and the frank hoods had a tuberous surface with small secondary tubercles (Figure 11.8A). The feebly marked hoods had a fine-grained surface without secondary tubercles (Figure 11.8C). The hoods were settled on the soft cushion and could easily be shifted. The sec-tion of the cushion showed it to be the jellylike mass. The hoods were clearly distinct from surrounding tissues.

At the age of two months, a slight thickening of the head skin and the appearance of weak tuberosity were visually observed in some fishes, but frank hoods were not detected. At the age of six months, 7.1% of fishes had frank hoods, 32.4% of fishes had feebly marked hoods and 60.6% of fishes had no hoods. At the ages of 14 and 24 months 20.2% and 43% of fishes

FIGURE 11.8 The macroscopic appearance of hoods. (A) frank hood; (B) big hood which completely covers the eye of the fish; (C) a feebly marked hood; (D) a fish without hood. *Source: [Kozlov et al., 2012], with permission.*

had frank hoods, respectively. These results of macroscopic observation indicate the progressive character of the hoods' appearance and growth and suggest their tumorous nature. The fact that these tumors were separate from adjacent tissues and did not connect to them is an indication that these were benign tumors. The headgrowths increased in size in older fishes. We observed a hood that completely covered an eye of one fish (Figure 11.8B). A hood that covered the whole surface of the head, including the gills and periocular region, at the age of two years was also described by Angelidis and co-authors [Angelidis et al., 2007]. Since this headgrowth disturbed the fish's eyesight and feeding, part of it was removed by surgery. During subsequent observation for 12 months recurrent growth of the hood was not observed, although recurrences of headgrowth after surgery were also reported [Angelidis et al., 2007].

Histological study of macroscopically normal head skin of goldfishes (Figure 11.9A) has shown that its structure corresponds to what has already been described in the literature. The fish skin consists of epidermis and derma. The epidermis is formed by Malpighian cells organized in several layers.

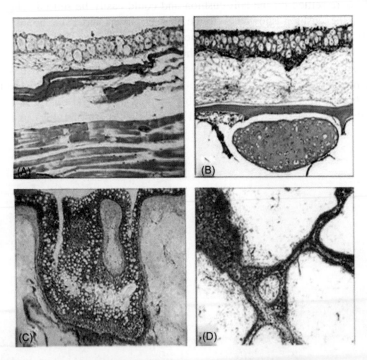

FIGURE 11.9 The histological dynamics of the hood development. (A) normal pavement epithelium; hemotoxylin-eosin, × 240; (B) epithelial hyperplasia at the early stage of tumor development; hemotoxylin-eosin, × 100; (C) submersible epithelial outgrowths with formation of flat inverted papilloma; hemotoxylin-eosin, × 180; (D) edema of the basal membrane with stroma overgrowth; PAS-reaction with diastase treatment, × 100. *Source: [Kozlov et al., 2012], with permission.*

In different layers, cells have different forms. In the lower layer, which is located on the basal membrane, the cells are cylindrical, in the medial layers they are polygonal and in upper layers they are flattened. The majority of authors consider fish skin epithelium to be multilayer pavement epithelium, although some authors consider it to be transitional epithelium [Ghadially and Whiteley, 1952]. Along with Malpighian cells, fish epidermis contains a considerable amount of mucus-producing cells. Mucus-producing cells are a peculiarity of fish skin. These cells, which are often called unicellular glands, secrete mucus consisting of mucopolysaccharides. The mucus forms a cuticle on the surface of the skin, protecting it from deleterious environmental factors. The epidermis also contains lymphocytes, macrophages and other cell types. The derma and hypoderma are localized under the epithelium. They contain pigment cells (chromatophores), mast cells and loose adipose tissue enriched with blood vessels.

In fishes 2 months and older, the microscopic studies detected irregular epithelial hyperplasia, the intensity of which increased with age. Epithelial overgrowth occurred generally due to an increase in the number of small cells resembling basal cells. This led to the formation of submersible outgrowths. The increase of mucus-producing cell size and accumulation of pigment under the basal membrane was also detected. Along with upsizing of the epidermal layer, thickening of the derma accompanied by edema took place (Figure 11.9B). With progression of the growth process, multiple deep submersible outgrowths merge with each other, creating the pattern characteristic of flat inverted papilloma. Further growth of the derma and an increase in mucus production with formation of intraepithelial cysts also took place (Figure 11.9C; Figure 11.10A,B).

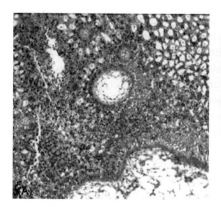
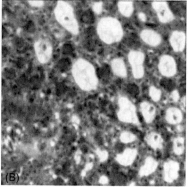

FIGURE 11.10   Solid structures of the hood. (A) the lamellar organization of tumor epithelium: ×100; (B) the part of figure 11.10A with greater magnification: tumor cells with large vacuoles, containing droplets of mucins; PAS-reaction with diastase treatment, ×200. *Source: [Kozlov et al., 2012], with permission.*

In some cases edema of the basal membrane was noted at the base of submersible outgrowths. Exophytic papillomatous formations with stroma overgrowth were also generated as a result of hyperplastic processes (Figure 11.9D).

A similar picture is presented in a few publications devoted to histological studies of goldfish hoods. Hyperplasia, disorganization and spongiosis of epidermis with desquamation of the surface layers, the increase in the number of mucus-producing cells in epithelium and chromatophores in derma and the increase of the volume of adipose tissue in the subepithelial layer were described [Angelidis et al., 2007].

Accumulation of pigment and mucus-producing cells with formation of cysts containing muscus was described for different hyperplastic processes and skin papilloma formation in hybrids of goldfishes [Lu et al., 2009], in different fish species [Steeves, 1969] and in virus-associated epidermal hyperplasia [Walker, 1969].

Headgrowths resembling hoods are described in *Bramble head* goldfishes, where they cover the surface of the head and gills in males. The microscopic studies of these headgrowths detected evident hyperplasia and submersible outgrowths with edema of stroma (the myxomatous degeneration) [Ghadially and Whiteley, 1952]. The authors consider that these headgrowths have a genetic and hormonal nature. They appear in males during the breeding season and afterwards undergo involution. They may also be caused by testosterone administration [Ghadially and Whiteley, 1952].

The hyperplastic outgrowths may also be caused by environmental factors. For example, papillomas of the skin and gills of aquarium fish in pulp mill effluents have been described [Lindesjoo and Thulin, 1994]. Papillomas of fish may be caused by exposure to chlorinated wastewater effluent [Grizzle et al., 1984].

Hyperplastic outgrowths in fish may be the result of viral infection. Cutaneous hyperplasia caused by *Herpesvirus cyprini* in cyprinids and walleye dermal sarcoma associated with epsilonretroviruses were discussed above in Part 3.1. Similar seasonal development of papillomas, containing virus-like particles, during winter and their regression in summer was described for hybrids of *Carassius auratus gibelio* and *Cyprinus carpio var. singuonensis* [Lu et al., 2009]. Virus-like particles were discovered in head skin papillomas of the Atlantic eel, *Anguilla vulgaris* [Deys, 1969].

Skin neoplasms, the majority of which are papillomas, are the most widespread in fishes [Deys, 1969; Steeves, 1969; Wellings, 1969; Groff, 2004]. In some fish species skin papillomas undergo malignization which is accompanied by the increase of mitotic activity in submersible outgrowths, loss of basal membrane and local invasion [Poulet et al., 1994; Syasin et al., 1999; Groff, 2004]. At the same time, in *Cyprinus carpio*, along with papilloma transformation into squamous cell carcinomas the regression and desquamation of papillomas were described [Wildgoose, 1992].

Besides skin papillomas, other kinds of tumors are described in goldfish including spindle-cell tumors resembling hemangiopericytoma [Morales and Schmidt, 1991], dermal fibrosarcoma with invasion of underlying tissues [Ahmed and Egusa, 1980] and nerve sheath tumors [Schlumberger, 1952].

As we discussed earlier in Part 3.1, fish tumors are generally less aggressive as compared with mammalian tumors. They are more differentiated and rarely exhibit metastatic behavior [Wellings, 1969; Noga, 2000; Roberts, 2001; Groff, 2004]. Attempts to transplant fish tumors are often unsuccessful [Wellings, 1969]. It is often difficult to distinguish tumor-like hyperplastic lesions of fish from true tumors [Groff, 2004]. Such lesions are called "pseudoneoplasms" [Harshbarger, 1984] or 'idiopathic epidermal proliferates" when they happen in skin [Noga, 2000]. These proliferates may simply be hyperplastic changes or may represent early stages of the tumor process [Noga, 2000]. It would be interesting to compare the hyperplastic overgrowths of fish skin with similar changes in rodents and humans.

Experimental application of carcinogens to rodent skin causes epithelial hyperplasia with formation of small local overgrowths and papillomas. The future of these tumors is diverse. They may stabilize in size, grow slowly, undergo involution or become malignant. Therefore experimentally observed hyperplastic changes may be considered as precancerous lesions [Shabad, 1967]. Atypical inflammatory epidermal overgrowths were repeatedly described in rabbits after subcutaneous injection of Scharlach R (scarlet red). They are characterized by evident epithelial hyperplasia and deep penetration of epidermal strands into derma that is reminiscent of the pattern of squamous cell carcinoma [Berenbein, 1980].

The interpretation of the nature of skin hyperplastic overgrowths in humans is also not simple. A.N. Apatenko [Apatenko, 1973] suggested discriminating among papillomatous malformations, reactive epithelial overgrowths, benign tumors, tumors with local destructive growth and malignant tumors (cancer). Malformations appear in early childhood without apparent cause. They represent exophytic outgrowths of connective tissue covered with multilayer pavement epithelium. Hyperkeratosis and acanthosis, formation of submersible outgrowths and proliferation of basal cells, in some cases are characteristic of this process. The growth of these papillomatous formations is very slow and malignization is observed extremely rarely. The histological structure of papillomas is close to that of malformations. Papillomas and malformations are distinguished only on the basis of clinical data about the time of their emergence and peculiarities of growth. Human benign skin tumors would include seborrheic warts, i.e. epithelial overgrowths protruding over the surface of the skin, with acanthosis and anastomosing strands [Apatenko, 1973]. The specific character of these tumors is hyperplasia of small epithelial cells similar to the basal cells described by us in fish. The formation of intraepidermal cysts and accumulation of melanin in the epithelial layer and in derma was also observed in seborrheic warts.

Examples of the regression of human skin tumors may be found in keratoacanthoma, which is defined as a benign tumor [Apatenko, 1973]. This tumor is characterized by evident hyperkeratosis of its central part and deep submersible growth of acanthotic strands into derma. In some cases it is difficult to morphologically discriminate this benign tumor from highly differentiated epidermoid cancer. But clinically keratoacanthoma has a tendency to spontaneous regression as opposed to cancer. Epidermal overgrowth with hyper- and parakeratosis, acanthosis and papillomatosis is observed in human senile skin keratoma, defined as precancerous tumor [Apatenko, 1973]. Histologocal studies of this tumor detect acanthotic strands located deep in the derma, and in some cases in polymorphic epithelium and disorganized basal membrane.

Reactive hyperplastic overgrowth of human epidermis, sometimes resembling cancer, is defined as "pseudocancer," or "pseudocarcinomatous hyperplasia" [Berenbein, 1980] (compare with "pseudoneoplasms" and "pseudodiseases" discussed in Parts 3.1 and 6.3, respectively). Acanthosis, submersible growth of epidermis, atypia, disorganization of basal membrane and deep penetration of strands into derma are described in pseudocancer. Morphological differential diagnostics of these changes and of epidermoid carcinoma is difficult. An indication of reactive changes may be the presence of inflammatory infiltration in derma [Apatenko, 1973; Berenbein, 1980]. The clinical observations show that pseudocancer has a tendency to involution after termination of the inflammatory process.

Keeping in the mind the comparative oncological analysis made above, the goldfish skin overgrowths described by us may be attributed to one of four kinds of changes: malformations, reactive proliferates, benign tumors or malignant tumors.

The hoods resemble malformations by the fact that they are genetically determined. At the same time these overgrowths differ from malformations by their progressive character of changes both at macroscopic and microscopic levels, which suggests their tumor nature. From reactive proliferates (pseudocancer), the hoods differ by the absence of inflammatory processes and the lack of tendency to regression. The hoods do not have the characteristics of malignancy such as atypia, mitoses and local invasion. Therefore, the most likely conclusion would be that the goldfish hoods represent genetically determined benign tumors [Kozlov et al., 2012].

11.2.2 The Goldfish Hood is an Artificially Selected Benign Tumor

Varieties of goldfish that develop headgrowths, or hoods (Oranda and Red cap Oranda, Lionhead, Ranchu), have been bred by selectionists during several hundred years. In this case, the fancy look of hoods on the heads of the goldfishes determined their selection and the successful growth of the population of fish endowed with this trait. According to our analysis these hoods are benign papillomas. That is, benign tumors were artificially selected for. To our knowledge, this is the first example of artificial selection of benign tumors described in the literature [Kozlov et al., 2012]. The possibility of artificial selection of tumors suggests that natural selection of tumors is also possible. We may infer that if artificial selection of tumors has worked in some cases, then natural selection might also use tumors as material for evolution. This conclusion supports the hypothesis of evolution by tumor neofunctionalization.

In ending this chapter, I wish to stress that our experimental data support, but do not prove, the hypothesis of evolution by tumor neofunctionalization. The phenomenon of *TSEEN* genes, or *carcino-evo-devo* genes, may have its own importance similar to that of carcinoembryonic antigens, independent of whether the hypothesis is right or wrong. The purpose of this chapter was not only to present supporting evidence obtained in my lab, but also to show that the hypothesis of evolution by tumor neofunctionalization makes nontrivial predictions that can be experimentally verified. This chapter outlines the possible directions of future research. I hope that it will stimulate many new experiments in many other laboratories.

Chapter 12

Other Evidence Supporting the Positive Evolutionary Role of Tumors and the Hypothesis of Evolution by Tumor Neofunctionalization

The whole content of this book, including our own experimental data, supports the positive evolutionary role of tumors and the hypothesis of evolution by tumor neofunctionalization. But there are also more data which I will briefly discuss in this chapter. Some of these data have been already mentioned above, but in a different context.

12.1 POSITIVE SELECTION OF MANY TUMOR-RELATED GENES IN PRIMATE LINEAGE

Several families of cancer/testis antigen genes including the *SPANX*, *GAGE* and *PRAME* gene families, with as-yet unknown functions, undergo positive selection in primate evolution [Kuprina et al., 2004b; Gjerstorff and Ditzel, 2008; Liu et al., 2008; Demuth and Hahn, 2009]. Rapid evolution of cancer/testis genes on the X chromosome has been demonstrated. Comparison of human:chimp orthologs of CT-X genes has shown that they diverge faster and undergo stronger positive selection than those on the autosomes or control genes on either X chromosomes or autosomes [Stevenson et al., 2007].

Mutations of the *BRCA1* gene in humans [Castilla et al., 1994; Friedman et al., 1994; Futreal et al., 1994; Miki et al., 1994; Hosking et al., 1995; Merajver et al., 1995a] and the wild type allele loss [Futreal et al., 1994; Cornelius et al., 1995; Merajver et al., 1995b] are associated with predisposition to breast and ovarian cancers and manifestation of the disease. *BRCA1* is considered to be a tumor suppressor gene [Xu et al., 2004]. Adaptive evolution of the tumor suppressor *BRCA1* in humans and chimps was demonstrated [Huttley et al., 2000]. Most of the internal *BRCA1* sequence is variable between primates and evolved under positive selection. BRCA1 C-terminal domains (BRCT) are highly conserved and are themselves not subject to positive selection [Pavlicek et al., 2004].

Angiogenin (ANG) is the tumor-growth promoter due to its ability to stimulate the formation of new blood vessels. Its expression is elevated in a variety of tumors. ANG is a member of the RNase A superfamily also known as RNase 5. The study among 11 primate species showed that the *ANG* gene has a significantly higher rate of nucleotide substitution at nonsynonymous sites than at synonymous sites, an indication of positive selection. The authors [Zhang and Rosenberg, 2002] called this "an intriguing phenomenon of unusual selective pressures on, and adaptive evolution of, cancer-related genes in primate evolution."

Other examples of positively selected tumor-related genes include Y-linked genes, some homeobox genes, centromeric histone genes, cadherins, cytochrome P450 genes, and genes in oncogenic viruses and are reviewed in [Crespi and Summers, 2006].

Analyzed alignments of 7645 chimp gene sequences to their human and mouse orthologs showed accelerated evolution in functions related to oncogenesis [Clark et al., 2003]. A scan for positively selected genes in the genomes of humans and chimps showed the evidence for positive selection in many genes involved in tumor suppression, apoptosis and cell cycle control [Nielsen et al., 2005].

Positive selection of many human tumor-related genes in the evolution of primates directly confirms the prediction of evolution by tumor neofunctionalization hypothesis concerning expression of evolutionarily new genes in tumors and selection for their new organismal function. If an evolutionarily new gene is expressed in tumors, or a sequence that is expressed in tumors acquires a function beneficial to the organism and becomes an evolutionarily new gene, selection of organisms for the enhancement of the new function should take place, as predicted by the hypothesis. This phenomenon, exactly, was found in research mentioned above: the references describe positive selection of genes and proteins in different primate groups, not the somatic evolution of tumor cells (see also the discussion in Part 10.8).

Positive selection of somatic mutations in human cancers also takes place [Babenko et al., 2006] and is easily explained. But this is not the case with the positive selection of many tumor-associated genes in the evolution of organisms, the paradox of which is difficult to explain other than by the postulation that tumors play a positive evolutionary role. The other attempt to explain positive selection of tumor-related genes is based on the concept of genomic conflict and antagonistic coevolution [Nielsen et al., 2005; Crespi and Summers, 2006]. See discussion of this concept in Part 10.2.

12.2 MORE EVOLUTIONARILY NOVEL GENES EXPRESSED IN TUMORS

The activation of processed pseudogenes in tumor cells [Moreau-Aubry et al., 2000; Zhang et al., 2003; Berger et al., 2005; Zheng et al., 2005] may be a step on the way to acquisition of a new function by the evolving gene.

Endoretroviral sequences which are expressed in tumors [Lower et al., 1993; Sauter et al., 1995; Anderson et al., 1998] have been shown to have originated as the result of retroviral infection of ancestral species and are relatively new for existing species of organisms. Some of them acquired new functions in the placenta which are beneficial to the organism [Volff, 2006].

Differential expression of human endogenous retroviruses (HERVs) is described in many human tumors: HERV-K family – in teratocarcinoma [Boller et al., 1993], in seminomas [Sauter et al., 1995], in breast cancer [Wang-Johanning et al., 2001], in urothelial and renal cell carcinomas [Florl et al., 1999], and in stomach cancers [Stauffer et al., 2004]; HERV-E – in prostate carcinoma [Wang-Johanning et al., 2003]; HERV-H – in leukemia cell lines [Lindeskog and Blomberg, 1997] and in cancers of the small intestine, bone marrow, bladder and cervix [Stauffer et al., 2004]. Although originally it was thought that HERVs are transcriptionally silent in most normal tissues [Florl et al., 1999], *in silico* data suggest that HERV-derived RNAs are more widely expressed in normal tissues than originally anticipated [Stauffer et al., 2004]. Indeed, *syncytin*, a domesticated retroviral gene that plays a role in placental biology, was identified in humans. This is the envelope gene of human defective endogenous retrovirus HERV-W. It is expressed in multinucleated placental syncytiotrophoblasts and may mediate placental cytotrophoblast fusion [Blond et al., 1999; Mi et al., 2000].

12.3 EXPRESSION OF MANY EVOLUTIONARILY NOVEL GENES IN THE PLACENTA, A TUMOR-LIKE ORGAN

The placenta is a tumor-like organ, as we realized in Parts 5.6 and 7.5. Its origin could be connected with retroviral infection in ancestor organisms which induced tumor growth and syncytiotrophoblast formation. *Syncytin* genes and *gag*-derived domesticated genes, which are essential for placenta formation [Volff, 2006], were evolutionarily new for the ancestral host organism when the retroviral infection happened that led to the placenta's origin. These evolutionarily novel genes were expressed in tumors induced by retroviruses and participated in the origin of placental function and the evolution of a new tumor-like organ.

There are also other evolutionarily novel genes expressed in the placenta. The study of the evolutionary origin of genes specifically expressed in the placenta according to microarray analysis showed that during the mature placental stages recently duplicated genes are predominantly expressed. In mice, genes expressed in the mature placenta are rodent specific and recently evolved. In humans, mature placental genes are primate specific. Thus newly evolved genes are expressed in mature placentas of both mice and humans. In mice, these genes include prolactine-like hormones, pregnancy-specific glycoproteins, carcinoembryonic antigen (CEA)-related cellular adhesion molecules and cathepsins. In humans, these genes code for pregnancy-specific glycoproteins and hormones [Knox and Baker, 2008].

Several genes expressed in the placenta, including the angiogenin gene (*ANG*), cadherin genes (*E-cadherin*, *P-cadherin* and *VE-cadherin*), the *pem* homeobox gene and various *ADAM* genes, evolve rapidly under positive selection (reviewed in [Crespi and Summers, 2006]).

12.4 ANTI-CANCER SELECTION MAY BE CONNECTED WITH DEVELOPMENTAL AND EVOLUTIONARY CONSTRAINTS

As it was discussed in Part 5.1, homeobox genes, the major development regulators, are involved in tumors. Several types of cancer in young children are associated with abnormalities in *Hox* gene expression and higher incidence of a cervical rib, i.e. a homeotic transformation of a cervical vertebra towards a thoracic-type vertebra [Galis, 1999]. Pediatric cancers are associated with a wide variety of congenital abnormalities, which suggests a link between morphogenesis and cancer risk, and connection of anti-cancer selection with developmental and evolutionary constraint [Galis and Metz, 2003; Kavanagh, 2003]. The evidence of such a connection supports the concept of a positive role of tumors in progressive evolution. As I discussed in Part 10.6, morphological novelties have been realized despite developmental constraints existing in the ancestor organisms. According to the tumor neofunctionalization hypothesis, tumors may represent a way to overcome the developmental constraints. Tumor-bearing organisms may represent transitional forms in the origin of evolutionary innovations.

Chapter 13

Overview

This book is about the possibility of the positive role of tumors in the evolution of multicellular organisms. Until now, the possibility of the positive evolutionary role of tumors was not addressed because different departments studied evolution and pathology. I believe that it was tumor processes, in particular heritable tumors, which provided evolving multicellular organisms with extra cell masses for the expression of evolutionarily novel genes, which originated in the germ plasm of evolving organisms, and for the construction of evolutionary innovations and morphological novelties. I formulate the hypothesis of evolution by tumor neofunctionalization, which I think is complementary to the hypothesis of evolution by gene duplication, and present supporting evidence in this book.

In this overview I will concentrate on the main arguments of the corresponding chapters without excessive detailing, and on their correlation with corresponding arguments in the other chapters. The references are not indicated here but can be found in the corresponding chapters.

1. The synthesis of evolutionary biology and the health sciences is emerging. New disciplines — Darwinian medicine, evolutionary epidemiology and evolutionary oncology — attempt to apply the evolutionary approach to their corresponding traditional areas of research.

Darwinian medicine focuses more on the individual patient whereas evolutionary epidemiology focuses on the spread of diseases. From the standpoint of Darwinian medicine, many diseases and health conditions have evolutionary origins. For example, senescence may be a result of the selection of traits that are advantageous at the early ages but are associated with adverse effects later in life. Evolutionary epidemiology assesses how traditional epidemiological characteristics such as lethality, illness, transmission rates, virulence and prevalence of infection change over time in the process of co-evolution of parasites and their hosts. The topic of this book is evolutionary oncology, i.e. the evolution of tumor-bearing organisms and the role of tumors in the evolution of organisms.

2. While Darwinian medicine and evolutionary epidemiology are looking for advantages that evolutionary biology could provide to the health sciences and medicine, evolutionary biology is interested in what role different pathologies could play in evolution. There are examples of pathogens and pathologies that have evolutionary importance. The most outstanding of those is the mutational process, which governs evolution, on one side, and generates various molecular diseases, on the other. I call this dualism the "evolution vs. pathology paradox" of mutations.

3. Tumors are widespread among multicellular organisms. Comparative oncology has generalized that neoplasia could be a property of all or most multicellular organisms but tumors are more frequent among the "higher" forms, e.g. insects and vertebrates, and in the evolutionarily more successful groups of organisms, e.g. in teleost fishes compared with cartilaginous fishes.

The methodology of neoplastic versus non-neoplastic differentiation may cause problems, and the line between hyperplasia and neoplasia could be uncertain. The term "pseudoneoplasms" was coined to designate non-neoplastic lesions in ectothermic animals that resemble neoplasms. Similar terminology ("pseudocancer" or "pseudocarcinomatous hyperplasia") is used for the designation of reactive hyperplastic overgrowth of human epidermis, sometimes resembling cancer. It is possible that borderline tumor-like processes exist. This possibility is important for considerations of what processes could supply evolving multicellular organisms with extra cell masses for the origin of new cell types, tissues and organs.

Neoplasms in fishes are generally less aggressive and more differentiated than neoplasms in mammals. They are mostly benign neoplasms. Asymptomatic human tumor lesions detected by early screening are less aggressive and grow more slowly, and the majority of individuals with such occult diseases will never develop the target cancer. Such lesions are called "pseudodiseases." The emerging picture is that the majority or at least a considerable part of tumors will never kill their hosts.

Ancient origin, occurrence in all multicellular organisms, and conservatism of cellular oncogenes and tumor suppressor genes support the concept that tumors are characteristic of all multicellular organisms and suggest that these genes have an important physiologic and evolutionary role.

The wide occurrence of tumors and tumor-like processes in multicellular organisms, tumors' connection to evolutionary success and progressive evolution, and the wide distribution and conservatism of cellular oncogenes suggest that tumors and/or some tumor-like processes could play a role in the evolution of multicellular organisms. That is, the "evolution vs. pathology paradox" may also exist for tumors.

4. There are features of tumors that could be used in evolution: many genes usually inactive in normal cells are activated in tumor cells; tumor cells can differentiate with concomitant loss of malignancy; tumors generate extra cell masses which are functionally not necessary for the organism and could be used as a building material for new cell types, tissues and organs; tumors have features of atypical tissues/organs and may participate in morphogenetic processes.

4.1. It has long been known that many tumors and cell lines derived from tumors are able to produce proteins which are not characteristic of the tissues and cell types of their origin. This phenomenon has been termed "ectopic syntheses." An example is polypeptide hormone production by neoplasms of non-endocrine tissues. The spectrum of ectopic hormones and the variety of hormone-producing tumors is impressive and includes dozens of different hormones and non-endocrine tumors.

The term "tumor cell plasticity" is currently used and implies the loss of specific gene expression and the concomitant aberrant expression of genes normally active in other cells. Such plasticity is especially pronounced in malignant tumor cells, as exemplified by vasculogenic mimicry manifested as tumor cell expression of the genes that are normally active in endothelial cells. Vasculogenic mimicry has been demonstrated in several tumor types.

Tumor antigens discovered in humans include oncofetal antigens, differentiation antigens, mutational antigens, overexpressed cellular antigens, viral antigens and cancer/testis antigens. It is currently believed that tumor-specific antigens are important for tumor immune surveillance. The priority cancer vaccine target antigens list includes several dozen cancer antigens, many of which have absolute tumor specificity.

Molecular markers of human malignant lesions have been described for detecting, localizing and monitoring the response of tumors to therapy. Tumor markers were originally defined as substances present at high levels in serum (i.e. serum markers) of patients with cancer and included oncofetal antigens, placental proteins, hormones, enzymes, catecholamine metabolites, polyamines, acute phase proteins and immunoglobulins.

Tumor-associated antigens of the cell surface and intracellular proteins specific to malignant cells currently are also represented among tumor markers, and they include not only protein-based markers but also RNA- and DNA-based markers. Tumor-associated antigens include viral antigens, MHC-related antigens, enzymes, oncogene products and cytogenetic markers.

Molecular hybridization approaches, DNA microarrays comparing global transcriptional profiles or gene expression signatures and other genome- and proteome-wide methodologies including computational differential display have led to the discovery of hundreds of genes expressed in tumors but not in corresponding normal tissues.

The accumulating data suggest that the fraction of the transcribed genome is greater than was earlier anticipated. A large portion of non coding regions in the human genome has been shown to be transcribed in tumors which results in many tumor-specific non coding RNAs.

We may conclude that many unusual genes, gene sets and non coding sequences may be activated in tumor cells. The activation of transcription in tumors is associated with epigenetic changes, i.e. changes in DNA methylation, histone modifications and changes in chromatin structure. This illustrates the magnitude of the biosynthetic potential of tumor cells that may be used in evolution. Novel genes and gene sets could be expressed in tumor cells and novel gene networks could be formed to participate in the origin of novel cell types, tissues and organs.

4.2. Another tumor property that might be used in evolution of organisms is the ability to differentiate with the loss of malignancy.

Tumor cells are characterized by deviation from typical differentiation, i.e., metaplasia, dysplasia and anaplasia. It is currently believed that this reflects the tumors' origins from undifferentiated stem cells. Nevertheless, in many malignant tumors some cells exhibit signs of aberrant differentiation that do not lead to normal histogenesis or organogenesis.

Induction of terminal differentiation *in vitro* with different agents has been accomplished with malignant cells derived from each of the three germ layers; with malignant cells originated from germ cells and embryonal tumors, and tumors from definitive tissues; with cells from hematopoietic malignancies and from solid tumors; and with cells from laboratory animals and from humans. Induction of terminal differentiation of tumor cells in culture has been achieved for squamous cell carcinoma, embryonal carcinoma, colon adenocarcinoma, breast adenocarcinoma, bladder transitional cell carcinoma, prostate cancer, melanoma, neuroblastoma, erythroleukemia and promyelocytic and myelocytic leukemias.

Cells from different tumors can also be induced by different agents to terminal differentiation *in vivo* in animal systems.

Tumors differ in potential for differentiation of their stem cells: embryonal carcinoma forms derivatives of all three germ layers while breast cancer stem cells form only glandular epithelium, etc.

Malignant cells are genetically abnormal. These abnormalities include aneuploidy, chromosome rearrangements, genetic changes in DNA, inactivation of suppressor genes, oncogenes activation etc. However, genetic abnormalities of tumor cells do not preclude the induction of tumor cell differentiation. Epigenetic mechanisms control gene expression in differentiation both during normal development and in induced differentiation of tumor cells. The suppression of malignancy in genetically abnormal tumor cells occurs due to epigenetic changes. "Epigenetics wins over genetics" in induction of differentiation of tumor cells.

The fact that agents inducing differentiation can do so with minimal effects on normal cells and at lower concentrations than cytotoxic drugs led to the emergence of an alternative approach to tumor cells killing by cytotoxic drugs, i.e. differentiation of cancer cells by changing the gene expression. The use of agents that can induce cancer cells' differentiation for therapeutic purposes in humans has been named "differentiation therapy."

So the available data indicate that malignancy is reversible. The genetic information responsible for the terminal differentiation of many types of tumor cells is present in these cells and may be expressed phenotypically. Tumor cells can differentiate with the loss of malignancy. The suppression of malignancy in tumor cells occurs due to epigenetic changes of gene expression. Genetic abnormalities of tumor cells do not preclude the induction of tumor cell differentiation. After tumor cells are induced to terminally differentiate, they stably lose their neoplastic growth capabilities.

4.3. Primitive multicellular organisms were small and had embryogenesis, which generated fixed amounts of cells with predetermined fates. The pre-existing cell types, tissues and organs of multicellular organisms have limited capacities for generation of new structures. The pre-existing cell types do not seem to be capable of providing the cell masses necessary for building new structures because of regulation and strict limitations imposed on the number of possible cell divisions.

The pre-existing cell types/tissues/organs already serve certain functions, which restricts their potential use for other purposes. The new structures with new functions could be generated only at the expense of the previous structures and functions, and examples of such economical use of pre-existing structures are known (e.g. derivatives of beetle elytra, mammalian auditory ossicles). Duplicated morphological structures may provide material for the origin of new structures, as in the case of the lower jaw of vertebrates which evolved from reiterated branchial arches of the agnathans.

The concept of "extra cell mass" exists in embryology as "set-aside cells" that do not participate in the embryonic specification and differentiation functions but have a function in the future development.

Asexual reproduction and regeneration are connected with the generation of considerable cell masses that participate in the formation of the missed "body-plan" structures. These processes may involve undifferentiated stem cells (neoblasts) in Planaria or dedifferentiated cells of the covering epithelium in Annelida. The question remains of whether these or similar processes, which are under control within the "body-plan" feedbacks and regulatory networks, can significantly contribute to the origin of evolutionarily new structures and organs.

As in the case of the origin of new genes from gene duplicates that are not functionally necessary to the organism, extra cell masses functionally unnecessary to the organism would be needed for the origin of evolutionarily new cell types, tissues and organs.

Tumors could provide such excessive cell masses to evolving multicellular organisms. The formation of excessive cell masses functionally unnecessary to the organism is considered a major feature of tumors, especially in invertebrates and lower vertebrates, where tumors can grow to considerable proportions.

4.4. Tumors have a hierarchical nature. They consist of a mixture of cell types at different stages of differentiation, similar to that in normal organs. At least two different populations of cells could be recognized: a minority of undifferentiated cancer stem cells (CSCs) and a majority of "derived cells," more differentiated and with a limited life span. The derived cells would include transit-amplifying cells and terminally differentiated cells. More cellular heterogeneity is added by different locations within a tumor mass and by the presence of non-tumor cells, i.e. inflammatory cells, tumor-associated fibroblasts, endothelial cells, pericytes, stem and progenitor cells of the tumor stroma and immature myeloid cells. Normal ancillary cells of different types play an important role in supporting the proliferation of tumor cells.

All this evidence suggests that tumors may be considered as complex heterogeneous tissues or organ-like systems or "quasi-tissues" that contain both stem cells and differentiated cells, or complex tridimensional tissues with phenotypically and functionally heterogeneous cells.

Even though the tumor cells are organized in organ-like structures, they still are not functionally necessary to the organism. Atypical tumor organs may be used by natural selection for the origin of normal organs if they acquire regulated functions and if the tumor-bearing organisms survive long enough to leave a progeny. The placenta, a tumor-like organ in eutherians, could be an example.

Tumors may participate in morphogenetic processes, as in the case of horny unpigmented papillomas or "cutaneous horns." Morphogenetic potential of tumors may be used in the origin of morphological novelties and morphological diversity.

The ability of tumor cells to differentiate with concomitant loss of malignancy, in combination with their ability to express genes that are not expressed in normal tissues and with other tumor features, may result in the emergence of new cell types, tissues and organs in evolution.

5, 5.1. Tumors might participate in the evolution of ontogenesis. According to Durante and Cohnheim's "embryonic remnants" theory of cancer, more embryonic cells are produced than are necessary for the formation of adult tissues and organs. The excess embryonic cells may stay dormant but capable of neoplastic development under appropriate conditions. Some modern authors think that the "embryonic remnants" concept is basically correct and there is considerable similarity of this concept with the modern theory of cancer stem cells.

The stem cell theory of cancer is the theory interrelating cancer, cell differentiation and embryonic development. This theory includes the notions of embryonic stem cells (ESCs), adult or somatic stem cells (SSCs) and cancer stem cells (CSCs).

Tumors contain a subpopulation of malignant cells with stem cell properties – CSCs. The cancer stem cell hypothesis suggests that tumorigenesis begins in normal adult stem cells or progenitor cells that have recently descended from them. Mutations in the DNA of normal stem cells, or epigenetic changes in normal stem cells, appear to be the initiating event in many cancers. CSCs may also arise from more restricted progenitor cells. The CSCs are similar to normal stem cells in their capability for self-renewal. They contain a multilineage differentiation capacity, although aberrant, and give rise to a hierarchy of progenitor and differentiated cells. CSCs also differ in many respects from the ESCs and SSCs. Most importantly, CSCs lack developmental control of ESCs and homeostatic control of SSCs populations. That is, tumors lack the functional feedback loop that regulates cancer stem cells.

CSCs cause tumors. They are often referred to as tumor-propagating cells (TPCs), reflecting their capability to cause the development of tumors when xenografted into immunodeficient mice. TPCs are the most tumorigenic cells of the tumor cell population. TPCs have defining markers. The function of CSCs' or TPCs' markers is not fully understood, but it is now recognized that some genes may cause tumors by activating an early developmental pathway associated with programming for multipotency, a feature of stem cells. The CSCs were physically isolated using the cell surface markers, first from leukemia and afterwards from solid tumors. After implantation of CSC-enriched populations, *in vivo* tumors were generated, with tumor cell populations no longer enriched for CSCs. That is, CSCs self-renew and also produce non-CSC, more differentiated progeny.

The signaling pathways that participate in regulation of self-renewal of normal SCs in different organs include Wnt, Sonic hedgehog (Shh), and Notch. These pathways participate in the normal developmental process. Wnt, Shh and Notch are among the fundamental signaling pathways that drive pattern formation in animals. But Wnt, Shh and Notch also participate in tumor progression when deregulated. Their deregulation causes deregulation of SCs self-renewal that leads to neoplastic growth. There are other genes, originally discovered as determinants of development, that demonstrate oncogenic potential.

On the other hand, many pathways classically associated with cancer may also regulate normal stem cell development. It is claimed that networks of protooncogenes and tumor suppressors have evolved to coordinately regulate stem cells. There are plenty of data on the connection of protooncogenes and tumor-suppressor genes with normal development. In fact, *Wnt*-1 was discovered as a protooncogene responsible for mammary tumors induced by the mouse mammary tumor virus. The *Wnt* gene family encodes a group of cell-signaling molecules that participate in vertebrates and invertebrates development.

The findings discussed above indicate the convergence of tumorigenic and embryonic signaling pathways. Canonical signaling pathways are fundamentally involved in the development of both normal organs and tumors of various histogenesis, i.e. atypical organs. Stem cells and tumor stem cells are regulated by common networks that include protooncogenes and tumor suppressors. This evidence suggests the fundamental connection of tumors and embryonic development and the possibility that tumors might participate in the evolution of ontogenesis.

5.2. Tumors are a disease of differentiation. Although CSCs have differentiation capabilities, they fail to differentiate correctly. This failure may lead to tumor formation. Tumor cells may be induced to differentiate experimentally, and patients with neoplasms could be treated with differentiation therapy, as discussed above. This means that the differentiation process is seriously impaired in tumors, and cancer in many respects may be considered as a disease of differentiation. Many authors have contributed to this concept.

The concept of tumors as a disease of differentiation may be summarized as follows: it is the overproduction of undifferentiated cells that causes tumors; tumors are the result of the loss/mutation of a differentiation product; and the loss of malignancy is connected with the presence of a differentiation product.

Thus the relationship between tumors and differentiation is dynamic. Tissue-specific functions are connected with their corresponding genes/differentiation products, and the loss of functions (e.g. due to mutations) is connected with tumors. The lack or disruption of the functional regulatory feedback loop may be the cause of tumors. On the other hand, the role of functional regulatory feedbacks appears as determinative in the differentiation of tumors when it occurs, and could be used in the evolutionary origin of new cell types/tissues/organs from tumors.

5.3. The epithelial to mesenchymal transition (EMT) occurs in normal and neoplastic development. EMT is the formation of mesenchymal cells from epithelia in different parts of an embryo or *in vitro* in epithelial tissue explants. EMT is the process of transdifferentiation, during which the epithelial phenotype transforms into a mesenchymal phenotype. EMT is not irreversible. The reverse process − mesenchymal to epithelial transition, through which mesenchymal cells re-establish epithelial phenotype − also occurs. Several rounds of EMT and MET take place during normal development. Thus EMT and MET play important roles in the differentiation and formation of many tissues and organs.

A wealth of data has accumulated proving that EMT and MET also participate in neoplastic development, i.e. in carcinoma progression and metastasis. Expression of many markers of mesenchymal phenotype has been described in a variety of neoplastic cell lines.

The same signaling pathways that regulate developmental EMT are also activated during tumor progression. EMT was demonstrated at the invasive front and in tumor cells migrating into stroma. On the other hand, most established metastases demonstrate an MET reversal.

Taken together, the different sets of data and concepts about stem cells, cancer as a disease of differentiation, signaling pathways and EMT provide an understanding of the fundamental and dynamic links between neoplastic and normal embryonic development. This fundamental dynamic linkage has existed throughout evolution and might have played a role in the evolution of ontogenesis. Throughout the evolution of multicellular organisms, their normal development was accompanied by neoplasms of different kinds, and evolution might use this connection for further elaboration of ontogenesis, or evo-devo.

5.4. Evolutionary biologists realized that it was the whole ontogenesis that evolved and was selected for fitness, not just the adult organisms. The outstanding generalizations on this relationship include von Baer's laws of development and Haeckel's biogenetic law. Severtsov defined the major modes of the evolution of ontogenesis, i.e. evolutionary changes of embryonic anlages and the addition of final stages of ontogenesis (anaboly).

The new synthesis between developmental biology and evolutionary biology is known as "evo-devo," or evolutionary developmental biology. Evolutionary developmental biology began using gene expression patterns to explain how higher taxa evolved. The remarkable homologies of homeobox genes, which control the anterior–posterior axis in vertebrates as well as in flies, and homologous developmental pathways have been discovered in numerous embryonic processes. Several major signaling pathways control the majority of events during the development of bilaterian animals, and possibly in all metazoans. The signaling pathways act by transcriptional activation of their target genes. Highly conserved genetic regulatory networks and regulatory sequences that play an important role in evolutionary developmental biology have been identified.

The correlation between rates of evolution of genes expressed during different stages of ontogenesis and evolutionary morphological conservation at different stages of ontogenesis was discovered both in animals and in plants. The slowest rates of gene evolution, which are associated with stronger negative selection, were found to be associated with the so-called "phylotypic stage" of embryogenesis. The phylotypic stage is characterized by the highest morphological similarity between embryos of different species belonging to the same phylum. On the other side, the highest rates of gene evolution (associated with the positive selection) are characteristic of genes expressed in adults, i.e. at terminal stages of ontogenesis.

Correlation of phylogenetic differences at later ontogenetic stages with the expression of newly evolved genes supports the role of novel genes in the origin of evolutionary innovations and morphological novelties. These data support von Baer's third law of development (earlier developmental stages are morphologically more similar across species than later stages).

Positive selection and accelerated evolution of many human tumor-related genes in primate lineage was described. This may indicate that heritable neoplasms, which are new to the host organism, may constitute an underlying basis for the emergence of evolutionary innovations and morphological novelties.

Embryonic tumors could play a role in the evolution of ontogenesis by evolutionary changes of embryonic anlages and by the addition of the final stages of development. The relatively late onset of adult cancers may be connected to the change of ontogenesis by means of addition of final stages of development, the anaboly modus of the evolution of ontogenesis described by Severtsov. If this assumption is correct, and neoplasms participated in anaboly, we may look for recapitulations of some tumor features in the most recently evolved organs.

5.5, 5.6. Indeed, the human brain, the most recently evolved human organ, recapitulates many features resembling those of tumors. The eutherian placenta is another evolutionary innovation that recapitulates many tumor features.

6. Tumors that might play a role in evolution include hereditary tumors; fetal, neonatal and infantile tumors; benign tumors, carcinomas *in situ* and pseudodiseases; tumors at the early and intermediate stages of progression; tumors that spontaneously regress through differentiation; and dormant tumors or tumors in a state of equilibrium.

However, malignant tumors at the late stages of progression cannot play such a role, although cases of spontaneous regression of malignant tumors are described and the placenta, a tumor-like organ in eutherians, has invasive potential.

7. There is a list of tumors that have played a role in evolution. This list includes the nitrogen-fixing root nodules of legumes; melanomatous cells that gave rise to macromelanophores in *Xiphophorus* fishes; benign tumors which, after artificial selection, gave rise to the hoods of goldfishes; malignant papillomatosis, which gave rise to symbiovilli in the stomach of voles; and the tumor-like organ, the eutherian placenta.

The "evolution vs. pathology paradox" may exist for tumors. Tumors could be considered as mutations at the multicellular level. On one hand, tumor processes may lead to various malignancies which may be harmful to individual organisms. On the other hand, tumor processes provide excess cell masses with high biosynthetic potential, and this may be used in progressive evolution for generation of new cell types, tissues and organs if organisms with tumors survive long enough to leave the progeny.

8. The hypothesis of evolution by tumor neofunctionalization suggests that expression of evolutionarily novel genes in tumors leads to the origin of new cell types, tissues and organs in progressive evolution.

The origin of novel genes is one of the major events in genome evolution. When the origin of evolutionarily novel genes is considered, it is implied that novel genes originate in the DNA of germ cells, so that they are inherited and later expressed in the individual development of the organism. The origin of new genes in somatic or tumor cells does not lead to their inheritance in progeny organisms and is not discussed in this book.

The phenomenon of the origin of novel genes is now well established. Novel genes may originate from pre-existing genes or *de novo*. It appears that gene duplication, exon shuffling, and *de novo* origin are the most fundamental modes, or principles, of the origin of novel genes. Transposons and endogenous retroviruses also participate in gene origin and genome evolution. There are many more molecular mechanisms that participate in the process of gene origin — several molecular mechanisms for each of the major modes of gene origin, and their combinations.

The general conclusion that authors make after considering different principles and mechanisms of the origin of novel genes is that this process is frequent, ongoing and widespread. The emerging picture is that genomes are plastic, evolving entities.

9. A concept that is now receiving more and more supporting evidence is that progressive evolution (i.e. the origin of morphological novelties and evolutionary innovations and the increase in complexity of multicellular organisms) is connected with the origin of evolutionarily novel genes, i.e. genes with altered or new functions as compared to pre-existing genes. This concept was elaborated by the collective effort of many scientists in the twentieth century.

9.1. Novel genes acquire new or altered functions. Different modes of gene origin — gene duplication, exon shuffling and *de novo* origin — are connected with somewhat different ways of acquiring new functions.

The evolutionary fate of duplicate genes can be pseudogenization, conservation and the increase of gene function (e.g. ribosomal proteins, rRNAs and histone genes); neofunctionalization, i.e. the origin of entirely new functions; and subfunctionalization when each of two genes retains only a subset of the original functions. The term "subneofunctionalization" designates situations when subfunctionalization is followed by neofunctionalization.

Exon shuffling evidently would lead to novel combinations of already existing functions. It played an important role in the evolution of the modular proteins involved in cell communications in Metazoa. Novel functions in Metazoa include those related to cell—cell interaction and adhesion. There are genes and proteins that participate in cell adhesion and cellular interactions that emerged in the deuterostome/chordate lineage and in vertebrates.

"Chimeric" or fusion genes evolve new functions because of considerable structural differences with their parental genes. In the evolution of chimeric fusion genes, rapid shifting away from ancestral functions (neofunctionalization) may be a general phenomenon.

De novo—originated genes should have some novel, previously nonexistent functions. These functions supposedly should fit into developmental or terminal differentiation classes of functions. This conclusion is in agreement with the different types of approaches that show that later ontogenetic stages are characterized by the expression of newly evolved genes.

The analysis of gene family evolution in yeast, flies and mammals has discovered common functional categories of rapidly evolving families, i.e. immune defense/stress response, metabolism, cell signaling, chemoreception and reproduction-related families. Gene families related to infection and disease demonstrate a consistent pattern of expansion. Genes with these functions also include the most rapidly evolving genes in *Drosophila* and mammals. In mammals, there is expansion of genes involved in olfaction, reproduction and immunity.

New genes quickly become essential, i.e. evolve essential functions and participate in development.

There are many examples of rapidly evolving novel genes supported by positive selection. These examples point out that novel genes and gene families are involved in adaptation and speciation. Positive Darwinian selection is supposed to be a driving force in the evolution of the novel genes in many cases, supporting the appearance of new adaptive functions.

Orphan genes are those which have no homologs in other species. Different genomes studied so far contain large numbers (10−20%) of orphan genes. Supposedly orphan genes may play species-specific adaptive roles.

Evolution at the level of DNA only is not sufficient for the emergence of a new function. It must be accompanied by expression of RNA and/or protein products and selection for their function in the organism. The existing evidence suggests the importance of the expression stage for the origin of new genes. Excessive tumor cells, functionally unnecessary to the organism, might be instrumental in facilitating the expression of evolutionarily novel genes and the origin of novel organismal gene functions.

9.2. The number of cell types in multicellular organisms has been increasing throughout evolution. The number of differentiated cell types in the multicellular organism is one of the major characteristics of its complexity. For example, humans have 411 different specialized cell types. Up to 479 morphological characters could be scored that may be an index of vertebrate morphological complexity. The original vertebrate innovations were first acquired on the chordate stem lineage (as compared to amphioxus and tunicates). Other innovative characters were acquired later along the jawed vertebrate stem lineage.

The diversity of domain combinations increases with the organism's complexity: the multidomain proteins and signaling pathways of the fly and worm are far more complex than those of yeast. Human genes are even more complex, and more alternative splicing events generate a larger number of protein products. The human proteome is more complex than that of invertebrates due to additional vertebrate-specific protein domains, a richer collection of domain architectures and a greater number of multidomain proteins.

Many authors associate evolutionary novelties and increase of morphological complexity in vertebrates with the increase of genome size and gene number due to gene duplication. Gene families expanded by gene duplication in vertebrates encode transcription factors, signaling molecules, structural proteins and enzymes. Evolution of different organs and the whole body plan is connected with the origin and expansion of corresponding gene families. Thus the greater number of *Hox* genes and their clusters, the increase in the general organization of the clusters and the global regulation enhancement were important for the growth of complexity and the adaptive evolution of vertebrates.

Many genes expressed in the neural crest are vertebrate innovations. These new genes originated either by gene divergence after duplication or by *de novo* gene evolution. They are connected with the appearance of novel molecular and cellular functions, and with tissues considered as vertebrate novelties.

Whole-genome duplications (WGDs) may be connected to an increase in morphological complexity. The two rounds of genome duplication in ancestors of vertebrates were followed by important morphological novelties such as enhanced nervous, endocrine and circulatory systems and sensory organs; more complex brains; the skull, vertebrae, endoskeleton and teeth; and in jawed vertebrates, paired appendages, hinged jaws and an adaptive immune system. Many authors believe that WGD events, although rare on an evolutionary scale, led to an increase in biological complexity and the origin of evolutionary innovations.

Correlation of phylogenetic differences at later ontogenetic stages with the expression of newly evolved genes supports the role of novel genes in the origin of evolutionary innovations and morphological novelties.

This discussion demonstrates that there is a connection between the origin of evolutionarily new genes and the emergence of evolutionary innovations and morphological novelties in multicellular organisms.

Scientists who study the origin of novel genes usually do not ask the question of where these genes will be expressed. The preconceived idea is that there are enough conditions for the new genes to be expressed in preexisting cells, or that extra cells will be automatically generated in organisms in adequate amounts for the expression of new genes and their novel functions. Nevertheless, many authors since August Weismann have pointed out the existence of competitive relationships and relationships of incompatibility between the genes which will disturb such a scenario. The expression of novel genes in the same cell may have energy constraints and selective disadvantages.

Scientists who assume that there is a general correlation between the increase in the gene number in the genomes of evolving organisms, from one side, and the increase in the number of cell types, the origin of other innovations and the overall complexity, on the other, should explain how such adequate correlation was realized at the cellular level. An adequate increase in cell number that accompanied the process of the origin of novel genes is hard to imagine. More likely, some autonomous cellular proliferative processes (tumors) were recruited to provide the space for the expression of new genes. New cell types may originate as a result of the coincidence of relatively independent processes at different structural levels (i.e. genes, cells and cell masses) and the "freezing" of the valuable coincidences by subsequent selection.

So tumors, similarly to gene duplication and retrotransposition events in the case of genome evolution, might provide the "raw material" for the origin of evolutionary innovations and morphological novelties.

10. This line of reasoning brings us to the hypothesis of evolution by tumor neofunctionalization, which may be formulated as follows:

Tumors are the source of extra cell masses which may be used in evolution of multicellular organisms for the expression of evolutionarily novel genes, for the origin of new differentiated cell types with novel functions and for building new structures which constitute evolutionary innovations and morphological novelties.

Hereditary tumors may play an evolutionary role by providing the conditions (space and resources) for the expression of genes newly evolving in the DNA of germ cells. As a result of the expression of novel genes, tumor cells may acquire new functions and differentiate in new directions which may lead to the origin of new cell types, tissues and organs. The new cell types are inherited in progeny generations due to genetic and epigenetic mechanisms similar to those for pre-existing cell types.

In addition to the discussions in Chapters 1–9, the following aspects were carefully examined as part of this hypothesis: gene competition as related to the possible evolutionary role of tumors; the possible evolutionary role of cellular oncogenes; the origin of feedback loops regulating new functions, new gene expression and new cell type proliferation; the inheritance of the new cell type in progeny generations; populations of tumor-bearing organisms with inherited tumors as transitional forms between established species of organisms at different stages of progressive evolution; the theory of "frozen accident" as related to the origin of new cell types/tissues/organs; the role of expression in the origin of new gene functions; tumors and the early evolution of Metazoa; the need for a continuous supply of the extra cell masses in the evolution of multicellular organisms; nonadaptive origins of organismal complexity; morphological novelties, developmental constraints and tumors; and cancer as a search engine for innovations and novel molecular combinations.

This examination has shown that the scenario described by the hypothesis of evolution by tumor neofunctionalization is feasible in terms of the possible mechanisms of its realization and in relation to other biological theories.

Among different aspects of the hypothesis, the most important question is the inheritance of the new cell type in the progeny generations. The inheritance of the new cell type is determined by the origin of evolutionarily novel genes in the germ plasm DNA; by heritability of tumors in which novel genes are expressed; and by epigenetic inheritance of gene expression patterns. So, the new cell type is inherited due to genetic and epigenetic mechanisms similar to those for pre-existing cell types.

11. As we can see from the above discussions, the hypothesis of evolution by tumor neofunctionalization is capable of satisfactorily describing the interrelated issues of individual, neoplastic and evolutionary development of multicellular organisms in relation to genome evolution and the origin of evolutionarily novel genes. But the hypothesis of evolution by tumor neofunctionalization makes it possible not only to explain and systematize a vast volume of observational and experimental data but also to formulate nontrivial predictions that can be experimentally tested. I made two strong predictions: (1) Evolutionarily new genes and/or sequences are specifically expressed in tumor cells, and (2) Tumors might be selected for new organismal functions.

To study experimentally the prediction concerning the expression of evolutionarily young or novel genes in tumors, two complementary approaches could be used. One is to study the evolutionary novelty of genes/sequences with proven tumor specificity of expression. The other is to study tumor specificity of expression of genes/sequences with proven evolutionary novelty. We performed both approaches and obtained similar results that support the hypothesis of evolution by tumor neofunctionalization.

We performed an analysis of the evolutionary novelty of tumor-specifically expressed EST sequences and showed that many of them are either evolutionarily new or relatively young and evolve neutrally. We have shown that *ELFN1-AS1* − a human gene with possible microRNA function expressed predominantly in tumors − originated in primates. We have shown that *PBOV1*, a human gene of recent *de novo* origin, has a highly tumor-specific expression profile. We also have demonstrated the evolutionary novelty of the whole class of human cancer/testis antigen genes. The data suggest that expression of proto-genes, evolutionarily young and/or novel genes in tumors may be considered as a new biological phenomenon, the phenomenon of tumor-specifically expressed, evolutionarily novel (*TSEEN*) or *carcino-evo-devo* genes.

Within the framework of the second strong prediction, we demonstrated that the goldfish hood is a benign tumor, which means that the tumor was artificially selected by selectionists over several hundred years.

So our own experimental studies confirmed the predictions made by the hypothesis of evolution by tumor neofunctionalization, but they do not prove this hypothesis. The phenomenon of *TSEEN* genes, or *carcino-evo-devo* genes, may have its own importance, similar to that of carcinoembryonic antigens, independent of whether the hypothesis is right or wrong. This chapter shows that the hypothesis of evolution by tumor neofunctionalization makes nontrivial predictions that can be experimentally verified and outlines the possible directions of future research.

12. The whole content of this book, including our own experimental data, supports the positive evolutionary role of tumors and the hypothesis of evolution by tumor neofunctionalization. The additional supporting lines of evidence include positive selection of many tumor-related genes in primate lineage; differential expression of human endogenous retroviruses (HERVs) in many human tumors; expression of evolutionarily novel and positively selected genes in the placenta, a tumor-like organ; and evidence of the connection of anti-cancer selection with developmental and evolutionary constraints that supports the concept of the positive role of tumors in progressive evolution.

Chapter 14

Conclusion

We see that tumors might play a positive role in evolution by providing evolving multicellular organisms with extra cell masses for the expression of evolutionarily new genes, for the origin of new differentiated cell types with novel functions and for building new structures that constitute morphological novelties and evolutionary innovations. Populations of tumor-bearing organisms could be transitional forms between established species of organisms at different stages of progressive evolution.

The hypothesis of evolution by tumor neofunctionalization explains the possible mechanisms of the origin of new cell types/tissues/organs in multicellular organisms that cannot be simplistically reduced only to the origin of new genes. This hypothesis explains the paradox of positive selection of tumor-related genes in the primate lineage, which is difficult to explain in any other way. It also explains how evolving organisms overcome ancestral developmental constraints and acquire morphological novelties. The hypothesis makes it possible to formulate nontrivial predictions, to plan new experiments, and to obtain original data.

The theory of the positive evolutionary role of tumors as a new scientific paradigm may considerably expand our understanding of the nature of tumors and of possibilities for influencing tumor processes. The realization that tumors are part of individual and evolutionary development and that a dormant or equilibrium state may be more typical for tumors than malignancy suggests a new strategy of keeping tumors in controlled chronic condition, contrary to the current strategy of killing tumors by all possible means, which has, in the long run, proven to be efficient only for certain types of tumors. Tumor balancing between equilibrium and malignancy supposes more delicate interventions which may move the balance towards an equilibrium state. Differentiation therapy and metronomic/adaptive therapy may be considered as first steps on this path. In this regard, expression of evolutionarily new genes in tumors may become the target of new therapeutic and/or preventive interventions. Approaches aimed at differentiation of tumor cells in new directions, obtaining new organismal functions by tumors and tumor incorporation into functional networks of the organism may become real in the near future.

Evolution by Tumor Neofunctionalization.

References

Abate-Shen, C., 2002. Deregulated homeobox gene expression in cancer: cause or consequence? Nat. Rev. Cancer 2, 777—785.

Abdelrazeg, A.S., 2007. Spontaneous regression of colorectal cancer: a review of cases from 1900 to 2005. Int. J. Colorectal Dis. 22, 727—736.

Abelev, G.I., Perova, S.D., Khramkova, N.I., et al., 1963a. Production of embryonal alpha-globulin by transplantable mouse hepatomas. Transplantation 1, 174—180.

Abelev, G.I., Perova, S.D., Khramkova, N.I., et al., 1963b. Embryonic serum alpha-globulin and its synthesis by transplantable mouse hepatoma. Biochimia. 28, 625—634.

Aburomia, R., Khaner, O., Sidow, A., 2003. Functional evolution in the ancestral lineage of vertebrates or when genomic complexity was wagging its morphological tail. J. Struct. Func. Genomics 3, 45—52.

Acevedo, H.F., Tong, J.Y., Hartsock, R.J., 1995. Human chorionic gonadotropin-β subunit gene expression in cultured human fetal and cancer cells of different types and origins. Cancer 76, 1467—1475.

Agrawal, N., Frederick, M.J., Pickering, C.R., et al., 2011. Exome sequencing of head and neck squamous cell carcinoma reveals inactivating mutations in *NOTCH1* (31 authors). Science 333, 1154—1157.

Aguirre-Ghiso, J.A., 2007. Models, mechanisms and clinical evidence for cancer dormancy. Nat. Rev. Cancer 7, 834—846.

Aharoni, A., Gaidukov, L., Khersonsky, O., et al., 2005. The "evolvability" of promiscuous protein functions. Nat. Genet. 37, 73—76.

Ahmed, A.T.A., Egusa, S., 1980. Dermal fibrosarcoma in goldfish *Carassius auratus* (L.). J. Fish Dis. 3, 249—254.

Ahuja, M.R., Anders, F., 1976. A genetic concept of the origin of cancer, based in part upon studies of neoplasms in fishes. Prog. Exp. Tumor Res. 20, 380—397.

Ajiki, W., Tsukuma, H., Oshima, A., et al., 1998. Effects of mass screening for neuroblastoma on incidence, mortality, and survival rates in Osaka, Japan. Cancer Causes Control 9, 631—636.

Akers, S.N., Odunsi, K., Karpf, A.R., 2010. Regulation of cancer germline antigen gene expression: implications for cancer immunotherapy. Future Oncol. 6, 717—732.

Akiva, P., Toporik, A., Edelheit, S., et al., 2006. Transcription-mediated gene fusion in the human genome. Genome Res. 16, 30—36.

Al-Hajj, M., Wicha, M.S., Benito-Hernandez, A., et al., 2003. Prospective identification of tumorigenic breast cancer cells. Proc. Natl. Acad. Sci. USA 100, 3983—3988.

Alba, M.M., Castresana, J., 2005. Inverse relationship between evolutionary rate and age of mammalian genes. Mol. Biol. Evol. 22, 598—606.

Alba, M.M., Castresana, J., 2007. On homology searches by protein Blast and the characterization of the age of genes. BMC Evol. Biol. 7, 53. Available from: http://dx.doi.org/10.1186/1471-2148-7-53.

Alema, S., Casalbore, P., Agostini, E., et al., 1985. Differentiation of PC12 phaeochromocytoma cells induced by v-src oncogene. Nature 316, 557—559.

Alexander, R.McN., 2006. Dinosaur biomechanics. Proc. Roy. Soc. B Available from: http://dx. doi.org/10.1098/rspb.2006.3532.

Alexandrova, S.A., Shvemberger, I.N., 2005. Genetic variability of the mouse hepatoma cells MH-22a revealed by RAPD-PCR-fingerprinting under different conditions of cultivation. Exp. Oncol. 27, 114–119.

Ali, F., Meier, R., 2008. Positive selection in *ASPM* is correlated with cerebral cortex evolution across primates but not with whole-brain size. Mol. Biol. Evol. 25, 2247–2250.

Alison, M.R., Islam, S., Wright, N.A., 2010. Stem cells in cancer: instigators and propagators? J. Cell Sci. 123, 2357–2368.

Allen, D.M., van Praag, H., Ray, J., et al., 2001. Ataxia telangiectasia mutated is essential during adult neurogenesis. Genes & Dev. 15, 554–566.

Allin, E.F., 1975. Evolution of mammalian middle ear. J. Morphol. 147, 403–437.

Allison, A.C., 1961. Genetic factors in resistance to malaria. Ann. N.Y. Acad. Sci. 91, 710–729.

Alonso, C.R., Wilkins, A.S., 2005. The molecular elements that underlie developmental evolution. Nat. Rev. Genet. 6, 709–715.

An, G., Ng, A.Y., Meka, C.S.R., et al., 2000. Cloning and characterization UROC28, a novel gene overexpressed in prostate, breast and bladder cancers. Cancer Res. 60, 7014–7020.

Anders, F., 1991. Contributions of Gordon-Kosswig melanoma system to the present concept of neoplasia. Pigm. Cell Res. 4, 7–29.

Anders, F., Schartl, M., Barnekow, A., et al., 1984. Xiphophorus as an in vivo model for studies on normal and defective control of oncogenes. Adv. Cancer Res. 42, 191–275.

Anderson, N.G., 1970. Evolutionary significance of virus infection. Nature 227, 1346–1347.

Anderson, C.R., Jensen, E.O., Llewellyn, D.J., et al., 1996. A new hemoglobin gene from soybean: a role for hemoglobin in all plants. Proc. Natl. Acad. Sci. USA 93, 5682–5687.

Anderson, A., Svensson, A., Rolny, C., et al., 1998. Expression of human retrovirus ERV3 (HERV-R) mRNA in normal and neoplastic tissues. Int. J. Oncol. 12, 309–313.

Andreef, M., Goodrich, D.W., Pardee, A.B., 2000. Cell proliferation, differentiation, and apoptosis. In: Bast, R.C., Kufe, D.W., Pollock, R.E., Weichselbaum, R.R., Holland, J.F., Frei., E. (Eds.), Holland-Frei Cancer Medicine, fifth ed. BC Decker, Hamilton (ON).

Angelidis, P., Vatsos, N.I., Karagiannis, D., 2007. Surgical excision of skin folds from the head of a goldfish *Carassius auratus* (Linnaeus 1758). J. Hellenic Vet. Med. Soc. 58, 299–305.

Anisimov, V.N., Ukraintseva, S.V., Yashin, A.I., 2005. Cancer in rodents: does it tell us about cancer in humans? Nat. Rev. Cancer 5, 807–819.

Apatenko, A.N., 1973. Epithelial Tumors and Malformations of Skin. Medicine, Moscow, 240 p.

Arechaga, J., 1993. On the boundary between development and neoplasia. An interview with Professor G. Barry Pierce. Int. J. Dev. Biol. 37, 5–16.

Arenas-Mena, C., Martinez, P., Cameron, R.A., et al., 1998. Expression of the *Hox* gene complex in the indirect development of a sea urchin. Proc. Natl. Acad. Sci. USA 95, 13062–13067.

Artieri, C.G., Haerty, W., Singh, R.S., 2009. Ontogeny and phylogeny: molecular signatures of selection, constraint, and temporal pleiotropy in the development of *Drosophila*. BMC Biol. 7, 42. Available from: http://dx.doi.org/10.1186/1741-7007-7-42.

Artiery, C.G., Singh, R.S., 2010. Demystifying phenotypes: the comparative genomics of evo-devo. Fly 4, 18–20.

Asashima, M., Komazaki, S., Satou, C., et al., 1982. Seasonal and geographical changes in spontaneous skin papillomas in the Japanese newt *Cynops pyrrhogaster*. Cancer Res. 42, 3741–3746.

Asashima, M., Oinuma, T., Matsutama, H., et al., 1985. Effects of temperature on papilloma growth in the newt, *Cynops pyrrhogaster*. Cancer Res. 45, 1198–1205.

Assis, R., Kondrashov, A.S., 2009. Rapid repetitive element-mediated expansion of piRNA clusters in mammalian evolution. Proc. Natl. Acad. Sci. USA 106, 7079–7082.

Atchley, W.R., Fitch, W.M., 1995. Myc and Max: molecular evolution of a family of proto-oncogene products and their dimerization partner. Proc. Natl. Acad. Sci. USA 92, 10217–10221.

Azumi, J.I., Sachs, L., 1977. Chromosome mapping of the genes that control differentiation and malignancy in myeloid leukemic cells. Proc. Natl. Acad. Sci. USA 74, 253–257.

Babenko, V.N., Basu, M.K., Kondrashov, F.A., et al., 2006. Signs of positive selection of somatic mutations in human cancers detected by EST sequence analysis. BMC Cancer 6, 36. Available from: http://dx.doi.org/10.1186/1471-2407-6-36.

Babushok, D.V., Ostertag, E.M., Kazazian Jr., H.H., 2006. Current topics in genome evolution: Molecular mechanisms of new gene formation. Cell. Mol. Life Sci. Available from: http://dx.doi.org/10.1007/s00018-006-6453-4.

Babushok, D.V., Ohshima, K., Ostertag, E.M., et al., 2007. A novel testis ubiquitin-binding protein gene arose by exon shuffling in hominoids. Genome Res. 17, 1129–1138.

Bacus, S.S., Stancovski, I., Huberman, E., et al., 1992. Tumor-inhibitory monoclonal antibodies to the HER-2/Neu receptor induce differentiation of human cancer cells. Cancer Res. 52, 2580–2589.

Bailey, J.A., Gu, Z., Clark, R.A., et al., 2002. Recent segmental duplications in the human genome. Science 297, 1003–1007.

Baillie, J.K., Barnett, M.W., Upton, K.R., et al., 2011. Somatic retrotransposition alters the genetic landscape of the human brain (19 authors). Nature 479, 534–537.

Baker, C.V.H., Bronner-Fraser, M., 1997. The origins of the neural crest. Part II: an evolutionary perspective. Mech. Dev. 69, 13–29.

Bakewell, M.A., Shi, P., Zhang, J., 2007. More genes underwent positive selection in chimpanzee evolution than in human evolution. Proc. Natl. Acad. Sci. USA 104, 7489–7494.

Balint, K., Xiao, M., Pinnix, C.C., et al., 2005. Activation of Notch1 signaling is required for β-catenin-mediated human primary melanoma progression. J. Clin. Invest. 115, 3166–3166.

Banyani, L., Varadi, A., Patthy, L., 1983. Common evolutionary origin of the fibrin-binding structures of fibronectin and tissue-type plasminogen activator. FEBS Lett. 163, 37–41.

Baranova, A.V., Lobashev, A.V., Ivanov, D.V., et al., 2001. In silico screening for tumor-specific expressed sequences in human genome. FEBS Lett. 508, 143–148.

Barber, B.J., 2004. Neoplastic diseases of commercially important marine bivalves. Aquat. Living Resour. 17, 449–466.

Barolo, S., Posakony, J.W., 2002. Three habits of highly effective signaling pathways: principles of transcriptional control by developmental cell signaling. Genes & Dev. 16, 1167–1181.

Bates, T.C., Luciano, M., Lind, P.A., et al., 2008. Recently-derived variants of brain-size genes *ASPM, MCPH1, CDK5RAP* and *BRCA1* not associated with general cognition, reading or language. Intelligence 36, 689–693.

Baylin, S.B., Herman, J.G., Graff, J.R., et al., 1998. Alterations in DNA methylation: a fundamental aspect of neoplasia. Adv. Cancer Res. 72, 141–196.

Beachy, P.A., Karhadkar, S.S., Berman, D.M., 2004. Tissue repair and stem cell renewal in carcinogenesis. Nature 432, 324–331.

Beard, J., 1902. Embryological aspects and etiology of carcinoma. Lancet 1, 1758–1761.

Beckwith, J.B., Perrin, E.V., 1963. In situ neuroblastomas: a contribution to the natural history of neural crest tumors. Amer. J. Path. 43, 1089–1104.

Begun, D.J., Lindfors, H.A., Thompson, M.E., et al., 2006. Recently evolved genes identified from *Drosophila yakuba* and *D. erecta* accessory gland expression sequence tags. Genetics 172, 1675–1681.

Begun, D.J., Lindfors, H.A., Kern, A.D., et al., 2007. Evidence for de novo evolution of testis-expressed genes in the *Drosophila yakuba/Drosophila erecta* clade. Genetics 176, 1131–1137.

Belancio, V.P., Hedges, D.J., Deininger, P., 2008. Mammalian non-LTR retrotransposons: for better or for worse, in sickness and in health. Genome Res. 18, 343–358.

Bell, G., Mooers, A.O., 1997. Size and complexity among multicellular organisms. Biol. J. Linn. Soc. 60, 345–363.

Bence-Jones, H., 1847. Papers on chemical pathology. Lecture III. Lancet 2, 269–272.

Benirschke, K., Garner, F.M., Jones, T.C. (Eds.), 1978. Pathology of Laboratory Animals. Springer-Verlag, New York, Heidelberg, Berlin.

Berdasco, M., Esteller, M., 2010. Aberrant epigenetic landscape in cancer: how cellular identity goes awry. Dev. Cell 19, 698–711.

Berenbein, B.A., 1980. Pseudocancer of Skin. Medicine, Moscow, 274 p.

Berezikov, E., Thuemmler, F., van Laake, L.W., et al., 2006. Diversity of microRNAs in human and chimpanzee brain. Nat. Genet. 38, 1375–1377.

Berger, I.R., Buschbeck, M., Bange, J., et al., 2005. Identification of a transcriptionally active hVH-5 pseudogene on 10q22.2. Cancer Genet. Cytogenet. 159, 155–159.

Berger, A.H., Knudson, A.G., Pandolfi, P.P., 2011. A continuum model for tumor suppression. Nature 476, 163–169.

Bergthorsson, U., Andersson, D.I., Roth, J.R., 2007. Ohno's dilemma: evolution of new genes under continuous selection. Proc. Natl. Acad. Sci. USA 104, 17004–17009.

Beringer, J.E., Brevin, N., Johnston, A.W.B., et al., 1979. The rhizobium–legume symbiosis. Proc. Roy. Soc. Lond. (Biol.) 204, 219–233.

Bernards, R., Weinberg, R.A., 2002. A progression puzzle. Nature 418, 823.

Bersaglieri, T., Sabeti, P.C., Patterson, N., et al., 2004. Genetic signatures of strong positive selection at the lactase gene. Am. J. Hum. Genet. 74, 1111–1120.

Betran, E., Long, M., 2002. Expansion of genome coding regions by acquisition of new genes. Genetica 115, 65–80.

Betran, E., Long, M., 2003. *Dntf-2r*, a young Drosophila retroposed gene with specific male expression under positive Darwinian selection. Genetics 164, 977–988.

Betran, E., Thornton, K., Long, M., 2002. Retroposed new genes out of the X in Drosophila. Genome Res. 12, 1854–1859.

Betran, E., Wang, W., Jin, L., et al., 2002b. Evolution of the phosphoglycerate mutase processed gene in human and chimpanzee revealing the origin of a new primate gene. Mol. Biol. Evol. 19, 654–663.

Bir, A.S., Fora, A.A., Levea, C., et al., 2009. Spontaneous regression of colorectal cancer metastatic to retroperitoneal lymph nodes. Anticancer Res. 29, 465–468.

Birney, E., Stamatoyannopoulos, J.A., Dutta, A., et al., 2007. Identification and analysis of functional elements in 1% of the human genome by the ENCODE pilot project. Nature 447, 799–816.

Birtle, Z., Goodstadt, L., Ponting, C., 2005. Duplication and positive selection among hominin-specific *PRAME* genes. BMC Genomics 6, 120. Available from: http://dx.doi.org/10.1186/1471-2164-6-120.

Bishop, J.M., 1983. Cellular oncogenes and retroviruses. Annu. Rev. Biochem. 52, 301–354.

Bittner, M., Meltzer, P., Chen, Y., et al., 2000. Molecular classification of cutaneous malignant melanoma by gene expression profiling. Nature 406, 536–540.

Blaise, S., de Parseval, N., Benit, L., et al., 2003. Genomewide screening for fusogenic human endogenous retrovirus envelopes identifies *syncytin 2*, a gene conserved in primate evolution. Proc. Natl. Acad. Sci. USA 100, 13013–13018.

Blond, J.-L., Beseme, F., Duret, L., et al., 1999. Molecular characterization and placental expression of HERV-W, a new human endogenous retrovirus family. J. Virol. 73, 1175–1185.

Bloom, J.D., Romero, P.A., Lu, Z., et al., 2007. Neutral genetic drift can alter promiscuous protein functions, potentially aiding functional evolution. Biol. Direct 2, 17. Available from: http://dx.doi.org/10.1186/1745-6150-2-17.

Boland, C.R., Goel, A., 2005. Somatic evolution of cancer cells. Semin. Cancer Biol. 15, 436–450.

Bolande, R.P., 1967. Cellular Aspects of Developmental Pathology. Lea and Fabirger, Philadelphia, pp. 88–143.

Bolande, R.P., 1971. Benignity of neonatal tumors and concept of cancer repression in early life. Amer. J. Dis. Child 122, 12–14.

Boller, K., Konig, H., Sauter, M., et al., 1993. Evidence that HERV-K is the endogenous retrovirus sequence that codes for the human teratocarcinoma-derived retrovirus HTDV. Virology 196, 349–353.

Boman, B.M., Wicha, M.S., 2008. Cancer stem cells: a step towards the cure. J. Clin. Oncol. 26, 2795–2799.

Bonasio, R., Tu, S., Reinberg, D., 2010. Molecular signals of epigenetic states. Science 330, 612–616.

Bonavia, E., 1895. Studies in the Evolution of Animals. London.

Bond, J., Roberts, E., Mochida, G.H., et al., 2002. *ASPM* is a major determinant of cerebral cortical size. Nat. Genet. 32, 316–320.

Bonnet, D., Dick, J.E., 1997. Human acute myeloid leukemia is organized as a hierarchy that originates from a primitive hematopoietic cell. Nat. Med. 3, 730–737.

Bordenstein, S.R., Marshall, M.L., Fry, A.J., et al., 2006. The tripartite associations between bacteriophage, Wolbachia, and arthropods. PLoS Pathog. 2, e43.

Bostwick, D.G., Cooner, W.H., Denis, L., et al., 1992. The association of benign prostatic hyperplasia and cancer of the prostate. Cancer 70, 291–301.

Boyd, W., 1966. The Spontaneous Regression of Cancer. Charles C. Thomas, Springfield, Ill.

Brabletz, T., Jung, A., Hermann, K., et al., 1998. Nuclear overexpression of the oncoprotein beta-catenin in colorectal cancer is localized predominantly at the invasion front. Pathol. Res. Pract. 194, 701–704.

Brabletz, T., Jung, A., Spaderna, S., et al., 2005. Migrating cancer stem cells: an integrated concept of tumor progression. Nat. Rev. Cancer 5, 744–749.

Bradley, J., Baltus, A., Skaletsky, H., et al., 2004. An X-to-autosome retrogene is required for spermatogenesis in mice. Nat. Genet. 36, 872–876.

Brault, V., Moore, R., Kutsch, S., et al., 2001. Inactivation of the β-catenin gene by Wnt1-Cre-mediated deletion results in dramatic brain malformation and failure of craniofacial development. Development 128, 1253–1264.

Braun, A.C., Wood, H.N., 1976. Suppression of the neoplastic state with the acquisition of specialized functions in cells, tissues and organs of crown gall teratomas of tobacco. Proc. Natl. Acad. Sci. USA 73, 496–500.

Braunstein, G.D., Vaitukaitis, J.L., Carbone, P.P., et al., 1973. Ectopic production of human chorionic gonadotrophin by neoplasms. Ann. Intern. Med. 78, 39–45.

Brawand, D., Soumillon, M., Necsulea, A., et al., 2011. The evolution of gene expression levels in mammalian organs (18 authors). Nature 478, 343–348.

Breitman, T.R., Selonick, S.E., Collins, S.J., 1980. Induction of differentiation of the human promyelocytic leukemia cell line (HL60) by retinoic acid. Proc. Natl. Acad. Sci. USA 77, 2936–2940.

Breitman, T.R., Collins, S.J., Keene, B.R., 1981. Terminal differentiation of human promyelocytic leukemic cells in primary culture in response to retinoic acid. Blood 57, 1000–1004.

Brennecke, J., Malone, C.D., Aravin, A.A., et al., 2008. An epigenetic role for maternally inherited piRNAs in transposon silencing. Science 322, 1387–1392.

Breslow, N., Chan, C.W., Dhom, G., et al., 1977. Latent carcinoma of prostate at autopsy in seven areas. Collaborative study organized by the International Agency for Research on Cancer, Lyon, France. Int. J. Cancer 20, 680–688.

Bridges, C.B., 1936. The Bar "gene" a duplication. Science 83, 210–211.

Brinster, R.L., 1974. The effect of cells transferred into mouse blastocyst on subsequent development. J. Exp. Med. 140, 1049–1056.

Britten, R.J., 2002. Divergence between samples of chimpanzee and human DNA sequences is 5%, counting indels. Proc. Natl. Acad. Sci. USA 99, 13633–13635.

Britten, R.J., 2004. Coding sequences of functioning human genes derived entirely from mobile element sequences. Proc. Natl. Acad. Sci. USA 101, 16825–16830.

Broders, A.C., 1932. Carcinoma in situ contrasted with benign penetrating epithelium. J.A.M.A. 99, 1670–1674.

Brosius, J., 1999. RNAs from all categories generate retrosequences that may be exapted as novel genes or regulatory elements. Gene 238, 115–134.

Brown, W.H., 1928. A case of pluriglandular syndrome: "Diabetes of bearded women." Lancet 2, 1022–1023.

Buckingham, M., Meilhac, S., Zaffran, S., 2005. Building the mammalian heart from two sources of myocardial cells. Nat. Rev. Genet. 6, 826–835.

Budd, G.E., Jensen, S., 2000. A critical reappraisal of the fossil record of the bilaterian phyla. Biol. Rev. 75, 253–295.

Bugnion, E., 1875. Ein Fall von Sarcoma beim Fische. Deutsch. Z. Tiermed. Vergl. Path. 1, 132–134.

Bustamante, C.D., Fledel-Alon, A., Williamson, S., et al., 2005. Natural selection on protein-coding genes in the human genome. Nature 437, 1153–1157.

Byrne, K.P., Wolfe, K.H., 2007. Consistent patterns of rate asymmetry and gene loss indicate widespread neofunctionalization of yeast genes after whole-genome duplication. Genetics 175, 1341–1350.

Byun-McKay, S.A., Geeta, R., 2007. Protein subcellular relocalization: a new perspective on the origin of novel genes. Trends Ecol. Evol. 22, 338–344.

Caballero, O.L., Chen, Y.-T., 2009. Cancer/testis (CT) antigens: potential targets for immunotherapy. Cancer Sci. 100, 2014–2021.

Cadigan, K.M., Nusse, R., 1997. Wnt signaling: a common theme in animal development. Genes & Dev. 11, 3286–3305.

Cai, J., Zhao, R., Jiang, H., 2008. De novo origination of a new protein-coding gene in Saccharomyces cerevisiae. Genetics 179, 487–496.

Cai, J.J., Petrov, D.A., 2010. Relaxed purifying selection and possibly high rate of adaptation in primate lineage-specific genes. Genome Biol. Evol. 2, 393–409.

Cairns, J., 1975. Mutation, selection and the natural history of cancer. Nature 255, 197–200.

Cairns, P., Sidransky, D., 1999. Molecular methods for diagnosis of cancer. Biochim. Biophys. Acta. 1423, C11–C18.

Callinan, P.A., Batzer, M.A., 2006. Retrotransposable elements and human disease. Genome Dyn. 1, 104–115.

Campbell Marotta, L.L., Polyak, K., 2009. Cancer stem cells: a model in the making. Curr. Opin. Genet. Develop. 19, 44–50.

Care, A., Silvani, A., Meccia, E., et al., 1996. HOXB7 constitutively activates basic fibroblast growth factor in melanomas. Mol. Cell. Biol. 16, 4842–4851.

Carninci, P., 2006. Tagging mammalian transcription complexity. Trends Genet. 22, 501–510.

Carninci, P., 2010. RNA dust: where are the genes? DNA Res. 17, 51–59.

Carroll, S.B., 1995. Homeotic genes and the evolution of arthropods and chordates. Nature 376, 479–485.

Carroll, S.B., 2001. Chance and necessity: the evolution of morphological complexity and diversity. Nature 409, 1102–1109.

Carroll, S.B., 2005. Evolution at two levels: on genes and form. PLoS Biol. 3, e245.

Carroll, S.B., 2008. Evo-devo and an expanding evolutionary synthesis: a genetic theory of morphological evolution. Cell 134, 25–36.

Carswell, E.A., Old, L.J., Kassel, R.L., et al., 1975. An endotoxin-induced serum factor that causes necrosis of tumors. Proc. Natl. Acad. Sci. USA 72, 3666–3670.

Carvunis, A.-R., Rolland, T., Wapinski, I., et al., 2012. Proto-genes and *de novo* gene birth (16 authors). Nature 487, 370–374.

Castilla, L.H., Cough, F.J., Erdos, M.R., et al., 1994. Mutations in the *BRCA1* gene families with early-onset breast and ovarian cancer. Nat. Genet. 8, 387–391.

Castillo-Davis, C.I., Hartl, D.L., 2002. Genome evolution and developmental constraint in *Caenorhabditis elegans*. Mol. Biol. Evol. 19, 728–735.

Cerling, T.E., Wynn, J.G., Andanje, S.A., et al., 2011. Woody cover and hominin environments in the past 6 million years. Nature 476, 51–56.

Chaffer, C.L., Weinberg, R.A., 2011. A perspective on cancer metastasis. Science 331, 1559–1564.

Chan, E.F., Gat, U., McNiff, J.M., et al., 1999. A common human skin tumor is caused by activating mutations in β-catenin. Nat. Genet. 21, 410–413.

Chang, T.-C., Yang, Y., Yasue, H., et al., 2011. The expansion of the *PRAME* gene family in Eutheria. PLoS ONE 6, e16867. Available from: http://dx.doi.org/10.1371/journal.pone.0016867.

Charafe-Jauffret, E., Monville, F., Ginestier, C., et al., 2008. Cancer stem cells in breast: current opinion and future challenges. Pathobiology 75, 75–84.

Charvet, C.J., Striedter, G.F., Finlay, B.L., 2011. Evo-devo and brain scaling: candidate developmental mechanisms for variation and constancy in vertebrate brain evolution. Brain Behav. Evol. 78, 248–257.

Chatterjee, S.K., Zetter, B.R., 2005. Cancer biomarkers: knowing the present and predicting the future. Future Oncol. 1, 37–50.

Cheever, M.A., Allison, J.P., Ferris, A.S., 2009. The prioritization of cancer antigens: a National Cancer Institute pilot project for the acceleration of translational research. Clin. Cancer Res. 15, 5323–5337.

Chen, F.C., Li, W.H., 2001. Genomic divergences between humans and other hominoids and the effective population size of the common ancestor of humans and chimpanzees. Am. J. Hum. Genet. 68, 444–456.

Chen, L., DeVries, A.L., Cheng, C.-H.C., 1997. Evolution of antifreeze glycoprotein gene from a trypsinogen gene in Antarctic notothenioid fish. Proc. Natl. Acad. Sci. USA 94, 3811–3816.

Chen, D., Livne-bar, I., Vanderluit, J.L., et al., 2004. Cell-specific effects of RB or RB/p107 loss on retinal development implicate an intrinsically death-resistant cell-of-origin in retinoblastoma. Cancer Cell 5, 539–551.

Chen, J.M., Stenson, P.D., Cooper, D.N., et al., 2005. A systematic analysis of LINE-1 endonuclease-dependent retrotranspositional events causing human genetic disease. Hum. Genet. 117, 411–427.

Chen, Y.-T., Scanlan, M.J., Venditti, C.A., et al., 2005a. Identification of cancer/testis-antigen genes by massively parallel signature sequencing. Proc. Natl. Acad. Sci. USA 102, 7940–7945.

Chen, Y.-T., Iseli, C., Venditti, C.A., et al., 2006. Identification of a new cancer/testis gene family, *CT47*, among expressed multicopy genes on the human X chromosome. Genes Chrom. Cancer 45, 392–400.

Chen, S.T., Cheng, H.C., Barbash, D.A., et al., 2007. Evolution of *hydra*, a recently evolved testis-expressed gene with nine alternative first exons in *Drosophila melanogaster*. PLoS Genet. 3, e107.

Chen, Y.T., Hsu, M., Lee, P., et al., 2009. Cancer/testis antigen CT45: analysis of mRNA and protein expression in human cancer (11 authors). Int. J. Cancer 124, 2893–2898.

Chen, S., Zhang, Y.E., Long, M., 2010. New genes in *Drosophila* quickly become essential. Science 330, 1682–1685.

Chen, Y.T., Chadburn, A., Lee, P., et al., 2010a. Expression of cancer testis antigen CT45 in classical Hodgkin lymphoma and other B-cell lymphomas (10 authors). Proc. Natl. Acad. Sci. USA 107, 3093–3098.

Cheng, Y.-H., Wong, E.W.P., Cheng, C.Y., 2011. Cancer/testis (CT) antigens, carcinogenesis and spermatogenesis. Spermatogenesis 1, 209–220.

Chenn, A., Walsh, C.A., 2002. Regulation of cerebral cortical size by control of cell cycle exit in neural precursors. Science 297, 365–369.

Cheung, H.-H., Lee, T.-L., Rennert, O.M., et al., 2009. DNA methylation of cancer genome. Birth Def. Res. (C) 87, 335–350.

Chilton, M.-D., Drummond, M.H., Merlo, D.J., et al., 1977. Stable incorporation of plasmid DNA into higher plants cells: the molecular basis of crown gall tumorigenesis. Cell 11, 263–271.

Chomez, P., De Backer, O., Bertrand, M., et al., 2001. An overview of *MAGE* gene family with the identification of all human members of the family. Cancer Res. 61, 5544–5551.

Chun, J., Schaltz, D.G., 1999a. Rearranging views on neurogenesis: neuronal death in the absence of DNA end-joining proteins. Neuron 22, 7–10.

Chun, J., Schatz, D.G., 1999b. Developmental neurobiology: alternative ends for a familiar story? Curr. Biol. 9, R251–R253.

Ciccarelli, F.D., von Mering, C., Suyama, M., et al., 2005. Complex genomic rearrangements lead to novel primate gene function. Genome Res. 15, 343–351.

Cillo, C., Barba, P., Freschi, G., et al., 1992. *HOX* gene expression in normal and neoplastic human kidney. Int. J. Cancer 51, 892–897.

Cillo, C., Faiella, A., Cantile, M., et al., 1999. Homeobox genes in cancer. Exp. Cell Res. 248, 1–9.

Cillo, C., Cantile, M., Faiella, A., et al., 2001. Homeobox genes in normal and malignant cells. J. Cell. Physiol. 188, 161–169.

Clamp, M., Fry, B., Kamal, M., et al., 2007. Distinguishing protein-coding and noncoding genes in the human genome. Proc. Natl. Acad. Sci. USA 104, 19428–19433.

Clark, W.H., 1991. Tumour progression and the nature of cancer. Br. J. Cancer 64, 631–644.

Clark, A.G., Glanowski, S., Nielsen, R., et al., 2003. Inferring nonneutral evolution from human-chimp-mouse orthologous gene trios (17 authors). Science 302, 1960–1963.

Clarke, M.F., Dick, J.E., Dirks, P.B., et al., 2006. Cancer stem cells – perspectives on current status and future directions: AACR workshop on cancer stem cells. Cancer Res. 66, 9339–9344.

Coffey, D.S., 2001. Similarities of prostate and breast cancer: evolution, diet, and estrogens. Urology 57, 31–38.

Coggin, J.H., Anderson, N.G., 1974. Cancer, differentiation and embryonic antigens: some central problems. Adv. Cancer Res. 19, 105–165.

Cohnheim, J., 1877. Vorlesungen uber allgemein Pathologie. Hirschwald, Berlin.

Cohnheim, J., 1889. Lectures on General Pathology, vol. 2. The New Syndenham Society, London.

Colbourne, J.K., Pfrender, M.E., Gilbert, D., et al., 2011. The ecoresponsive genome of *Daphnia pulex* (69 authors). Science 331, 555–561.

Cole, W.H., 1974. Spontaneous regression of cancer: the metabolic triumph of the host? Ann. NY Acad. Sci. 230, 111–141.

Coley, W.B., 1893. The treatment of malignant tumors by repeated inoculations of erysipelas: with a report of ten original cases. Am. J. Med. Sci. 105, 487–511.

Coley, W.B., 1894. Treatment of inoperable malignant tumors with the toxins of erysipelas and the bacillus prodigiosus. Am. J. Med. Sci. 108, 50–66.

Collinge, W.E., 1891. Note on a tumor in *Anadonta cygnea* Linn. J. Anat. Physiol. Norm. Path. 25, 154.

Collins, S.J., 2008. Retinoic acid receptors, hematopoiesis and leukemogenesis. Curr. Opin. Hematol. 15, 346–351.

Collins, A.G., Valentine, J.W., 2001. Defining phyla: evolutionary pathways to metazoan body plans. Evol. Develop. 3, 432–442.

Collins, S.J., Ruscetti, F.W., Gallagher, R.E., et al., 1978. Terminal differentiation of human promyelocytic leukemia cells induced by dimethyl sulfoxide and other compounds. Proc. Natl. Acad. Sci. USA 75, 2458–2462.

Comeron, J.M., 1995. A method for estimating the numbers of synonymous and nonsynonymous substitutions per site. J. Mol. Evol. 41, 1152–1159.

Comeron, J.M., 1999. K-Estimator: calculation of the number of nucleotide substitutions per site and the confidence intervals. Bioinformatics 9, 763–764.

Conant, G.C., Wolfe, K.H., 2008. Turning a hobby into a job: how duplicated genes find new functions. Nat. Rev. Genet. 9, 938–950.

Connoly, J.L., Schnitt, S.J., Wang, H.H., et al., 2000. Principles of cancer pathology. In: Bast, R. C., Kufe, D.W., Pollock, R.E., Weichselbaum, R.R., Holland, J.F., Frei., E. (Eds.), Holland-Frei Cancer Medicine, fifth ed. BC Decker, Hamilton (ON).

Conway Morris, S., 1998. Early metazoan evolution: reconciling paleontology and molecular biology. Amer. Zool. 38, 867–877.

Copley, S.D., 2003. Enzymes with extra talents: moonlighting functions and catalytic promiscuity. Curr. Opin. Chem. Biol. 7, 265–272.

Cordain, L., Eaton, S.B., Sebastian, A., et al., 2005. Origin and evolution of the Western diet: health implications for the 21st century. Am. J. Clin. Nutr. 81, 341–354.

Cordaux, R., Batzer, M.A., 2009. The impact of retrotransposons on human genome evolution. Nat. Rev. Genet. 10, 691–703.

Cornelis, R.S., Neuhausen, S.L., Johansson, O., et al., the Breast Cancer Linkage Consortium 1995. High allele loss rates at 17q12-q21 in breast and ovarian tumors from *BRCA1*-linked families. Genes Chrom. Cancer 13, 203–210.

Cosgrove, G.E., O'Farrell, T.P., 1965. Papillomas and other lesions in the stomach of pine mice. J. Mammalogy 46, 510–513.

Coufal, N.G., Garcia-Perez, J.L., Peng, G.E., et al., 2009. L1 retrotransposition in human progenitor cells (10 authors). Nature 460, 1127–1131.

Count, E.W., 1947. Brain and body weight in man: their antecedents in growth and evolution. Ann. N.Y. Acad. Sci. 46, 993–1122.

Cox, R.L., Stephens, R.E., Reinisch, C.L., 2003. p63/73 homologues in surf clam: novel signaling motifs and implications for control of expression. Gene 320, 49–58.

Crespi, B., Summers, K., 2005. Evolutionary biology of cancer. Trends Ecol. Evol. 20, 545–552.

Crespi, B.J., Summers, K., 2006. Positive selection in the evolution of cancer. Biol. Rev. 81, 407–424.

Crick, F.H., 1968. The origin of the genetic code. J. Mol. Biol. 38, 367–379.

Cripps, R.M., Olson, E.N., 2002. Control of cardiac development by an evolutionarily conserved transcriptional network. Develop. Biol. 246, 14–28.

Crow, K.D., Stadler, P.F., Lynch, V.J., et al., 2006. The "fish-specific" Hox cluster duplication is coincident with the origin of teleosts. Mol. Biol. Evol. 23, 121–136.

Currat, M., Excoffier, L., Maddison, W., et al., 2006. Comment on "Ongoing adaptive evolution of *ASPM*, a brain size determinant in *Homo sapiens*" and "*Microcephalin*, a gene regulating brain size, continues to evolve adaptively in humans." Science 313, 172a.

Cutter, A.D., Ward, S., 2005. Sexual and temporal dynamics of molecular evolution in *C. elegans* development. Mol. Biol. Evol. 22, 178–188.

Cvekl, A., Piatigorsky, J., 1996. Lens development and crystalline gene expression: many roles for Pax-6. BioEssays 18, 621–630.

Daenen, S., Vellenga, E., van Dobbenburgh, O.A., et al., 1986. Retinoic acid as antileukemic therapy in a patient with acute promyelocytic leukemia and Aspergillus pneumonia. Blood 67, 559–561.

Dai, H., Chen, Y., Chen, S., et al., 2008. The evolution of courtship behaviors through the origination of a new gene in *Drosophila*. Proc. Natl. Acad. Sci. USA 105, 7478–7483.

Dalerba, P., Cho, R.W., Clarke, M.F., 2007. Cancer stem cells: models and concepts. Annu. Rev. Med. 58, 267–284.

Darby, J.F., Workman, P., 2011. Many faces of a cancer-supporting protein. Nature 478, 334–335.

Davidson, E.H., 1991. Spatial mechanisms of gene regulation in metazoan embryos. Development 113, 1–26.

Davidson, E.H., 2006. The Regulatory Genome. Elsevier, Academic Press, San Diego, USA.

Davidson, E.H., Peterson, K.J., Cameron, R.A., 1995. Origin of bilaterian body plans: evolution of developmental regulatory mechanisms. Science 270, 1319–1325.

Davis, J.C., Petrov, D.A., 2004. Preferential duplication of conserved proteins in eukaryotic genomes. PLoS Biol. 2, 0318–0326.

Davis, J.C., Brandman, O., Petrov, D.A., 2005. Protein evolution in the context of *Drosophila* development. J. Mol. Evol. 60, 774–785.

Dawe, C.J., 1969. Phylogeny and oncogeny. Natl. Cancer Inst. Monogr. 31, 1–41.

Dawe, C.J., 1973. Comparative neoplasia. In: Holland, J.F., Frei, E. (Eds.), Cancer Medicine. Lea and Febiger, Philadelphia, pp. 193–240.

Dawe, C.J., Harshbarger, J.C. (Eds.), 1969. Neoplasms and related disorders of invertebrate and lower vertebrate animals. Natl. Cancer Inst. Monogr. 31, 772.

De Robertis, E.M., 2008. Evo-Devo: variation on ancestral themes. Cell 132, 185–195.

Deininger, P.L., Batzer, M.A., 1999. *Alu* repeats and human disease. Mol. Genet. Metab. 67, 183–193.

Demuth, J.P., Hahn, M.W., 2009. The life and death of gene families. BioEssays 31, 29–39.

Demuth, J.P., De Bie, T., Stajich, J.E., et al., 2006. The evolution of mammalian gene families. PLoS ONE 1, e85. Available from: http://dx.doi.org/10.1371/journal.pone.0000085.

Deng, C., Cheng, C.-H.C., Ye, H., et al., 2010. Evolution of an antifreeze protein by neofunctionalization under escape from adaptive conflict. Proc. Natl. Acad. Sci. USA 107, 21593–21598.

de Rosa, R., Grenier, J.K., Andreeva, T., et al., 1999. Hox genes in brachiopods and priapulids and protostome evolution. Nature 399, 772–776.

de The, H., Chomienne, C., Lanotte, M., et al., 1990. The t(15;17) translocation of acute promyelocytic leukaemia fuses the retinoic acid receptor alpha gene to a novel transcribed locus. Nature 347, 558–561.

Dethlefsen, L., McFall-Ngai, M., Relman, D.A., 2007. An ecological and evolutionary perspective on human-microbe mutualism and disease. Nature 449, 811–818.

Devor, E.J., 2006. Primate microRNAs miR-220 and miR-492 lie within processed pseudogenes. J. Hered. 97, 186–190.

Dewell, R.A., 2000. Colonial origin for Eumetazoa: major morphological transitions and the origin of bilaterian complexity. J. Morph. 243, 35–74.

Deys, B.F., 1969. Papillomas in the Atlantic eel, *Anguilla vulgaris*. Natl. Cancer Inst. Monogr. 31, 187–193.

Di Cristofano, A., Pesce, B., Cordon-Cardo, C., 1998. *Pten* is essential for embryonic development and tumor suppression. Nat. Genet. 4, 348–355.

Dickinson, M.E., Krumlauf, R., McMahon, A.P., 1994. Evidence for a mitogenic effect of Wnt-1 in the developing mammalian central nervous system. Development 120, 1453–1471.

Dickinson, D.J., Nelson, W.J., Weis, W.I., 2011. A polarized epithelium organized by β- and α-catenin predates cadherin and metazoan origins. Science 331, 1336–1339.

Dietz, M., Fouchey, S., Furmanski, P., 1981. Spontaneous regression of Friend-virus-induced erythroleukemia. VII. The genetic control of regression. Int. J. Cancer 27, 341–348.

Djebali, S., Davis, C.A., Merkel, A., et al., 2012. Landscape of transcription in human cells (85 authors). Nature 489, 101–108.

Dobrynin, P., Matyunina, E., Malov, S.V., et al., 2013. The novelty of human cancer/testis antigen encoding genes in evolution. Int. J. Genomics 2013, 105108, http://dx.doi.org/10.1155/2013/105108.

Dobson-Stone, C., Gatt, J.M., Kuan, S.A., et al., 2007. Investigation of MCPH1 G37995C and ASPM A44871G polymorphisms and brain size in a healthy cohort. Neuroimage 37, 394–400.

Dobzhansky, T., 1973. Nothing in biology makes sense except in the light of evolution. Amer. Biol. Teacher 35, 125−129.

Dodueva, I.E., Frolova, N.V., Lutova, L.A., 2007. Plant tumorigenesis: different ways for shifting systemic control of plant cell division and differentiation. Transgenic Plant J. 1, 17−34.

Dollinger, M.R., Ratner, L.H., Shanoian, C.A., et al., 1967. Carcinoid syndrome associated with pancreatic tumors. Archs. Intern. Med. 120, 575−580.

Domazet-Loso, T., Tautz, D., 2003. An evolutionary analysis of orphan genes in *Drosophila*. Genome Res. 13, 2213−2219.

Domazet-Loso, T., Tautz, D., 2008. An ancient evolutionary origin of genes associated with human genetic diseases. Mol. Biol. Evol. 25, 2699−2707.

Domazet-Loso, T., Tautz, D., 2010a. Phylostratigraphic tracking of cancer genes suggests a link to the emergence of multicellularity in metazoan. BMC Biol. 8, 66.

Domazet-Loso, T., Tautz, D., 2010b. A phylogenetically based transcriptome age index mirrors ontogenetic divergence patterns. Nature 468, 815−819.

Domazet-Loso, T., Brajkovic, J., Tautz, D., 2007. A phylostratigraphy approach to uncover the genomic history of major adaptations in metazoan lineages. Trends Genet. 23, 533−539.

Donoghue, P.C.J., Graham, A., Kelsh, R.N., 2008. The origin and evolution of the neural crest. Bioessays 30, 530−541.

Doolittle, R.F., 1985. The genealogy of some recently evolved vertebrate proteins. Trends Biochem. Sci. 10, 233−237.

Dorit, R.L., Schoenbach, L., Gilbert, W., 1990. How big is the universe of exons? Science 250, 1377−1382.

Dorus, S., Vallender, E.J., Evans, P.D., et al., 2004. Accelerated evolution of nervous system genes in the origin of *Homo sapiens*. Cell 119, 1027−1040.

Duboule, D., 1995. Vertebrate *Hox* genes and proliferation: an alternative pathway to homeosis? Curr. Opin. Genet. Dev. 5, 525−528.

Duboule, D., 2007. The rise and fall of Hox gene clusters. Development 134, 2549−2560.

Dupressoir, A., Marceau, G., Vernochet, C., et al., 2005. Syncytin-A and syncytin-B, two fusogenic placenta-specific murine envelope genes of retroviral origin conserved in Muridae. Proc. Natl. Acad. Sci. USA 102, 725−730.

Dupressoir, A., Vernochet, C., Bawa, O., et al., 2009. Syncytin-A knockout mice demonstrate the critical role in placentation of a fusogenic, retrovirus-derived envelop gene. Proc. Natl. Acad. Sci. USA 106, 12127−12132.

Durante, F., 1874. Nesso fisio-patologico tra la struttura dei nei materni e la genesi di alcuni tumori maligni. Arch. Memor. Observ. Chir. Prat. 11, 217.

Duret, L., Mouchirod, D., 2000. Determinants of substitution rates in mammalian genes: expression pattern affects selection intensity but not mutation rate. Mol. Biol. Evol. 17, 68−74.

Duret, L., Chureau, C., Samain, S., et al., 2006. The *Xist* RNA gene evolved in eutherians by pseudogenization of a protein-coding gene. Science 312, 1653−1655.

Dvorak, H.F., 1986. Tumors: wounds that do not heal. Similarities between tumor stroma generation and wound healing. N. Engl. J. Med. 315, 1650−1659.

Dvorak, H.F., 2002. Vascular permeability factor/vascular endothelial growth factor: a critical cytokine in tumor angiogenesis and a potential target for diagnosis and therapy. J. Clin. Oncol. 20, 4368−4380.

Easton, D.F., Bishop, D.T., Ford, D., et al., the Breast Cancer Linkage Consortium 1993. Genetic linkage analysis in familial breast and ovarian cancer: results from 214 families. Am. J. Hum. Genet. 52, 678−701.

Eaton, S.B., Konner, M., Shostack, M., 1988. Stone agers in the fast lane: chronic degenerative diseases in evolutionary perspective. Amer. J. Med. 84, 739–749.

Effron, M., Griner, L., Benirschke, K., 1977. Nature and rate of neoplasia found in captive wild mammals, birds, and reptiles at necropsy. J. Natl. Cancer Inst. 59, 185–198.

Ehrlich, M., 2002. DNA methylation in cancer: too much, but also too little. Oncogene 21, 5400–5413.

Ehrlich, M., 2009. DNA hypomethylation in cancer cells. Epigenomics 1, 239–259.

Eichler, E.E., 2001. Recent duplication, domain accretion and the dynamic mutation of the human genome. Trends Genet. 17, 661–669.

El-Wakeel, H., Umpleby, H.C., 2003. Systematic review of fibroadenoma as a risk factor for breast cancer. Breast 12, 302–307.

Emerson, J.J., Kaessmann, H., Betran, E., et al., 2004. Extensive gene traffic on the mammalian X chromosome. Science 303, 537–540.

Emlen, D.J., Lavine, L.C., Ewen-Campen, B., 2007. On the origin and evolutionary diversification of beetle horns. Proc. Natl. Acad. Sci. USA 104, 8661–8668.

Emlen, D.J., Warren, I.A., Johns, A., et al., 2012. A mechanism of extreme growth and reliable signaling in sexually selected ornaments and weapons. Science 337, 860–864.

Endo, T., Ikeo, K., Gojobori, T., 1996. Large-scale search for genes on which positive selection may operate. Mol. Biol. Evol. 13, 685–690.

Erwin, D.H., Davidson, E.H., 2002. The last common bilaterian ancestor. Development 129, 3021–3032.

Esteller, M., 2007. Cancer epigenomics: DNA methylomes and histone-modification maps. Nat. Rev. Genet. 8, 286–298.

Evans, A.E., Gerson, J., Schnaufer, L., 1976. Spontaneous regression of neuroblastoma. Natl. Cancer Inst. Monogr. 44, 49–54.

Evans, P.D., Anderson, J.R., Vallender, E.J., et al., 2004a. Adaptive evolution of *ASPM*, a major determinant of cerebral cortical size in humans. Hum. Mol. Genet. 13, 489–494.

Evans, P.D., Anderson, J.R., Vallender, E.J., et al., 2004b. Reconstructing the evolutionary history of *microcephalin*, a gene controlling human brain size. Hum. Mol. Genet. 13, 1139–1145.

Evans, P.D., Gilbert, S.L., Mekel-Bobrob, N., et al., 2005. *Microcephalin*, a gene regulating brain size, continues to evolve adaptively in humans. Science 309, 1717–1722.

Evdokimova, V., Tognon, C., Ng, T., et al., 2009. Translational activation of Snail1 and other developmentally regulated transcription factors by YB-1 promotes an epithelial-mesenchymal transition. Cancer Cell 15, 402–415.

Everson, T.C., 1964. Spontaneous regression of cancer. Ann. N.Y. Acad. Sci. 114, 721–735.

Everson, T.C., Cole, W.H., 1956. Spontaneous regression of cancer: preliminary report. Annals of Surgery 144, 366–380.

Everson, T.C., Cole, W.H., 1966. Spontaneous Regression of Cancer. J.B. Saunders & Co., Philadelphia.

Evtushenko, V.I., Hanson, K.P., Barabitskaya, O.V., et al., 1989. Determination of the upper limit of rat genome expression. Mol. Biol. (Moscow) 23, 663–675.

Ewald, P.W., 1994. Evolution of Infectious Disease. Oxford University Press, Oxford, New York.

Eyler, C.E., Rich, J.N., 2008. Survival of the fittest: cancer stem cells in therapeutic resistance and angiogenesis. J. Clin. Oncol. 26, 2839–2845.

Eyre-Walker, A., Keightley, P.D., 1999. High genomic deleterious mutation rates in hominids. Nature 397, 344–347.

Eyre-Walker, A., Keightley, P.D., Smith, N.G.C., et al., 2002. Quantifying the slightly deleterious mutation model of molecular evolution. Mol. Biol. Evol. 19, 2142–2149.

Eyre-Walker, A., Woolfit, M., Phelps, T., 2006. The distribution of fitness effects of new deleterious amino acid mutations in humans. Genetics 173, 891–900.

Fan, X., Eberhart, C.G., 2008. Medulloblastoma stem cells. J. Clin. Oncol. 26, 2821–2827.

Faulkner, G.J., Kimura, Y., Daub, C.O., et al., 2009. The regulated retrotransposon transcriptome of mammalian cells (22 authors). Nat. Genet. 41, 563–571.

Fearon, E.R., Vogelstein, B., 2000. Tumor suppressor gene defects in human cancer. In: Bast, R.C., Kufe, D.W., Pollock, R.E., Weichselbaum, R.R., Holland, J.F., Frei., E. (Eds.), Holland-Frei Cancer Medicine, fifth ed. BC Decker, Hamilton (ON).

Feinberg, A.P., Vogelstein, B., 1983. Hypomethylation distinguishes genes of some human cancers from their normal counterparts. Nature 303, 89–92.

Feinberg, A.P., Ohlsson, R., Henikoff, S., 2005. The epigenetic progenitor origin of human cancer. Nat. Rev. Genet. 7, 21–33.

Feng, S., Jacobsen, S.E., Reik, W., et al., 2010. Epigenetic reprogramming in plant and animal development. Science 330, 622–627.

Fennell, R.H., 1955. Carcinoma in situ of the cervix with early invasive changes. Cancer 8, 302–309.

Fennell, R.H., Castleman, B., 1955. Carcinoma in situ. New Eng. J. Med. 252, 985–990.

Fernandez, A.A., Morris, M.R., 2008. Mate choice for more melanin as a mechanism to maintain functional oncogene. Proc. Natl. Acad. Sci. USA 105, 13503–13507.

Ferner, K., Mess, A., 2011. Evolution and development of fetal membranes and placentation in amniote vertebrates. Respir. Physiol. Neurobiol. 178, 39–50.

Feschotte, C., Pritham, E.J., 2007. DNA transposons and the evolution of eukaryotic genomes. Annu. Rev. Genet. 41, 331–368.

Fibach, E., Sachs, L., 1975. Control of normal differentiation of myeloid leukemic cells. VIII. Induction of differentiation to mature granulocytes in mass culture. J. Cell. Physiol. 86, 221–230.

Florl, A.R., Lower, R., Schmitz-Drager, B.J., et al., 1999. DNA methylation and expression of LINE-1 and HERV-K provirus sequences in urothelial and renal cell carcinomas. Br. J. Cancer 80, 1312–1321.

Force, A., Lynch, M., Picket, F.B., et al., 1999. Preservation of duplicate genes by complementary, degenerative mutations. Genetics 151, 1531–1545.

Forrester, L.M., Brunkow, M., Bernstein, A., 1992. Proto-oncogenes in mammalian development. Curr. Opin. Genet. Dev. 2, 38–44.

Fortna, A., Kim, Y., MacLaren, E., et al., 2004. Lineage-specific gene duplication and loss in humans and great ape evolution (16 authors). PLoS Biol. 2, 0937–0954.

Foulds, L., 1954. The experimental study of tumor progression: a review. Cancer Res. 14, 327–339.

Fox, H., 1912. Observations upon neoplasms in wild animals in the Philadelphia zoological gardens. J. Path. Bact. 17, 217–231.

Fraga, M.F., Ballestar, E., Villar-Garea, A., et al., 2005. Loss of acetylation at Lys16 and trimethylation at Lys20 of histone H4 is a common hallmark of human cancer. Nat. Genet. 37, 391–400.

Francino, M.P., 2005. An adaptive radiation model for the origin of new gene functions. Nat. Genet. 37, 573–577.

Frank, S.A., 2004. Genetic predisposition to cancer – insights from population genetics. Nat. Rev. Genet. 5, 764–772.

Fratta, E., Coral, S., Covre, A., et al., 2011. The biology of cancer testis antigens: putative function, regulation and therapeutic potential. Mol. Oncol. 5, 164–182.

Freeling, M., Thomas, B.C., 2006. Gene-balanced duplications, like tetraploidy, provide predictable drive to increase morphological complexity. Genome Res. 16, 805–814.

Friedman, M., Marshal-Jones, O., Ross, E.J., 1966. Cushing's syndrome: adrenocortical hyperactivity secondary to neoplasms outside the pituitary-adrenal system. Q. J. Med. 35, 196–244.

Friedman, L.S., Ostermeyer, E.A., Szabo, C.I., et al., 1994. Confirmation of *BRCA1* by analysis of germline mutations linked to breast and ovarian cancer in ten families. Nat. Genet. 8, 399–404.

Friend, C., Scher, W., Holland, J.G., et al., 1971. Hemoglobin synthesis in murine virus-induced leukemic cells in vitro: stimulation of erythroid differentiation by dimethyl sulfoxide. Proc. Natl. Acad. Sci. USA 68, 378–382.

Frohman, L.A., 1991. Endocrine manifestations of neoplasia. twelfth ed. In: Wilson, J.D., Braunwald, E., Isselbacher, K.J. et al., Harrison's Principles of Internal Medicine, vol. 2. McGraw Hill, New York, NY, pp. 1638–1641.

Fusco, F.D., Rosen, S.W., 1966. Gonadotropin-producing anaplastic large-cell carcinoma of the lung. New Engl. J. Med. 275, 507–515.

Futreal, P.A., Liu, Q.Y., Shattuck Evidence, D., et al., 1994. *BRCA1* mutations in primary breast and ovarian carcinomas. Science 266, 120–122.

Galachyants, Y., Kozlov, A.P., 2009. CDD as a tool for discovery of specifically-expressed transcripts. Russ. J. AIDS, Cancer Public Health 13 (2), 60–61, http://www.aidsconference.spb.ru/articles/arc9UuPPc.pdf.

Galis, F., 1999. Why do almost all mammals have seven cervical vertebrae? Developmental constraints, *Hox* genes, and cancer. J. Exp. Zool. (Mol. Dev. Evol.) 285, 19–26.

Galis, F., Metz, J.A.J., 2003. Anti-cancer selection as a source of developmental and evolutionary constraints. BioEssays 25, 1035–1039.

Gama-Sosa, M.A., Slagel, V.A., Trewyn, R.W., et al., 1983. The 5-methylcytosine content of DNA from human tumors. Nucleic Acids Res. 11, 6883–6894.

Gao, Y., Sun, Y., Frank, K.M., et al., 1998. A critical role for DNA end-joining proteins in both lymphogenesis and neurogenesis. Cell 95, 891–902.

Gao, Y., Ferguson, D.O., Xie, W., et al., 2000. Interplay of p53 and DNA-repair protein XRCC4 in tumorigenesis, genomic stability and development. Nature 404, 897–900.

Garcia-Fernandez, J., 2005. The genesis and evolution of homeobox gene clusters. Nat. Rev. Genet 6, 881–892.

Garfield, D.A., Wray, G.A., 2009. Comparative embryology without a microscope: using genomic approaches to understand the evolution of development. J. Biol. 8, 65. Available from: http://dx.doi.org/10.1186/jbiol161.

Gatenby, R.A., 2009. A change of strategy in the war on cancer. Nature 459, 508–509.

Gatenby, R.A., Silva, A.S., Gillies, R.J., et al., 2009. Adaptive therapy. Cancer Res. 69, 4894–4903.

Gaudet, F., Hodgson, J.G., Eden, A., et al., 2003. Induction of tumors in mice by genomic hypomethylation. Science 300, 489–492.

Gerhart, J., 1999. 1998 Warkany lecture: signalling pathways in development. Teratology 60, 226–239.

Gerschenson, M., Graves, K., Carson, S.D., et al., 1986. Regulation of melanoma by the embryonic skin. Proc. Natl. Acad. Sci. USA 83, 7307–7310.

Gestl, S.A., Leonard, T.L., Biddle, J.L., et al., 2007. Dormant Wnt-initiated mammary cancer can participate in reconstituting functional mammary glands. Mol. Cell Biol. 27, 195–207.

Getchell, R.G., Casey, J.W., Bowser, P.R., 1998. Seasonal occurrence of virally induced skin tumors in wild fish. J. Aquat. Anim. Health. 10, 191−201.

Ghadially, F.N., Whiteley, H.J., 1952. Hormonally induced epithelial hyperplasia in the goldfish (*Carassius auratus*). Brit. J. Cancer 6, 246−248.

Gibbons, A., 2009. What's for dinner? Researchers seek our ancestors' answers. Science 326, 1478−1479.

Gilbert, W., 1978. Why genes in pieces? Nature 271, 501.

Gilbert, S.F., Opiz, J.M., Raff, R.A., 1996. Resynthesizing evolutionary and developmental biology. Dev. Biol. 173, 357−372.

Gilbert, S.L., Dobyns, W.B., Lahn, B.T., 2005. Genetic links between brain development and brain evolution. Nuture Rev. Genet. 6, 581−590.

Gille, C., Polette, M., Piette, J., et al., 1966. Vimentin expression in cervical carcinomas: association with invasive and migratory potential. J. Pathol. 180, 175−180.

Gimm, O., Attie-Bitach, T., Lees, J.A., et al., 2000. Expression of the PTEN tumor suppressor protein during human development. Hum. Mol. Genet. 9, 1633−1639.

Giovannucci, E., 2003. Nutrition, insulin, insulin-like growth factors, and cancer. Horm. Metab. Res. 35, 694−704.

Gjerstorff, M.F., Ditzel, H.J., 2008. An overview of the GAGE cancer/testis antigen family with the inclusion of newly identified members. Tissue Antigens 71, 187−192.

Glazunov, M.F., 1947. Classification and nomenclature of tumors and tumor-like processes. In: Petrov, N.N. (Ed.), Malignant Tumors, vol. 1. Medgiz, Leningrad.

Globus, J., Kuhlenbeck, H., 1942. Tumors of the striatothalamic and related regions: their probable source of origin and more common forms. Arch. Pathol. (Chic) 24, 674−737.

Goelz, S.E., Vogelstein, B., Hamilton, S.A., et al., 1985. Hypomethylation of DNA from benign and malignant human colon neoplasms. Science 228, 187−190.

Gold, P., Freeman, S.O., 1965. Demonstration of tumor-specific antigens in human colonic carcinoma by immunological tolerance and absorption techniques. J. Exp. Med. 121, 439−445.

Goldberg, W.M., McNeil, M.J., 1967. Cushing's syndrome due to an ACTH-producing carcinoma of the thyroid. Can. Med. Ass. J. 96, 1577−1579.

Goldschmidt, R., 1940. The Material Basis of Evolution. Yale Univ. Press, New Haven, 436 pp.

Goodman, C.S., Coughlin, B.C., 2000. The evolution of evo-devo biology. Proc. Natl. Acad. Sci. USA 97, 4424−4425.

Goodson, M.S., Crookes-Goodson, W.J., Kimbell, J.R., et al., 2006. Characterization and role of p53 family members in the symbiont-induced morphogenesis of the *Euprymna scolopes* light organ. Biol. Bull. 211, 7−17.

Gootwine, E., Webb, C.G., Sachs, L., 1982. Participation of myeloid leukaemic cells injected in embryos in hematopoietic differentiation in adult mice. Nature 299, 63−65.

Gordon, M., 1927. The genetics of viviparous top-minnow Platypoecilus: the inheritance of two kinds of melanophores. Genetics 12, 253−283.

Gould, S.J., 1977. Ontogeny and Phylogeny. Harvard Univ. Press, Cambridge.

Gowen, L.C., Johnson, B.L., Latour, A.M., et al., 1996. *Brca1* deficiency results in early embryonic lethality characterized by neuroepithelial abnormalities. Nat. Genet. 12, 191−194.

Graham, J., 1992. Cancer Selection. The New Theory of Evolution. Aculeus Press Inc., Lexington, 213 pp.

Green, H.S.N., 1940. Familial mammary tumors in the rabbit. IV. The evolution of autonomy in the course of tumor development as indicated by transplantation experiments. J. Exp. Med. 71, 305−324.

Gregory, W.K., 1951. Evolution Emerging. MacMillan, New York.

Grimaldi, D., Engel, M.S., 2005. Evolution of the Insects. Cambridge Univ. Press.

Grizzle, J.M., Goodwin, A.E., 1998. Neoplasms and Related Lesions. In: Leatherland, J.F., Woo, P.T.K., (Eds.), Fish Diseases and Disorders, vol. 2: Non-infectious Disorders, pp. 37−104.

Grizzle, J.M., Melius, P., Strength, D.R., 1984. Papillomas of fish exposed to chlorinated waste-water effluent. J. Natl. Cancer Inst. 73, 1133−1142.

Groff, J.M., 2004. Neoplasia in fishes. Vet. Clin. Exot. Anim. 7, 705−756.

Groszer, M., Erickson, R., Scripture-Adams, D.D., et al., 2001. Negative regulation of neural stem/progenitor cell proliferation by the *Pten* tumor suppressor gene in vivo. Science 294, 2186−2189.

Groszer, M., Erickson, R., Scripture-Adams, D.D., et al., 2006. PTEN negatively regulates neural stem self-renewal by modulating G_0-G_1 cell cycle entry. Proc. Natl. Acad. Sci. USA 103, 111−116.

Gu, Z., Nicolae, D., Lu, H.H-S., et al., 2002. Rapid divergence in expression between duplicate genes inferred from microarray data. Trends Genet. 18, 609−613.

Gupta, P.B., Chaffer, C.L., Weinberg, R.A., 2009. Cancer stem cells: mirage or reality? Nat. Med. 15, 1010−1012.

Guray, M., Sahin, A.A., 2006. Benign breast diseases: classification, diagnosis, and management. Oncologist 11, 435−449.

Gurchot, C., 1975. The trophoblast theory of cancer (John Beard, 1857 − 1924) revisited. Oncology 31, 310−333.

Gurchot, C., Krebs Jr., E.T., Krebs, E.T., 1947. Growth of human trophoblast in eye of rabbit. Its relationship to the origin of cancer. Surg. Gynec. Obstet. 84, 301−312.

Haas, P.G., Delongchamps, N., Brawley, W.O., et al., 2008. The worldwide epidemiology of prostate cancer: perspectives from autopsy studies. Can. J. Urol. 15, 3866−3871.

Haeckel, E., 1866. Generelle Morphologie der Organismen: Allgemeine Grundzüge der organischen Formen-Wissenschaft, mechanisch begründet durch die von Charles Darwin reformirte Descendenz-Theorie, 2 vols. Georg Reimer, Berlin, 574, pp. 462.

Hahn, M.W., 2007. Detecting natural selection on *cis*-regulatory DNA. Genetica 129, 7−18.

Hahn, M.W., 2008. Toward a selection theory of molecular evolution. Evolution 62, 255−265.

Hahn, M.W., 2009. Distinguishing among evolutionary models for the maintenance of gene duplicates. J. Hered. 100, 605−617.

Hahn, M.W., Wray, G.A., 2002. The g-value paradox. Evol. Dev. 4, 73−75.

Hahn, M.W., De Bie, T., Stajich, J.E., et al., 2005. Estimating the tempo and mode of gene family evolution from comparative genomic data. Genome Res. 15, 1153−1160.

Hahn, M.W., Demuth, J.P., Han, S.-G., 2007. Accelerated rate of gene gain and loss in primates. Genetics 177, 1941−1949.

Hahn, M.W., Han, M.V., Han, S.-G., 2007b. Gene family evolution across 12 *Drosophila* genomes. PLoS Genet. 3, e197. Available from: http://dx.doi.org/10.1371/journal.pgen.0030197.

Hakem, R., de la Pompa, J.L., Sirard, C., et al., 1996. The tumor suppressor gene *Brca1* is required for embryonic cellular proliferation in the mouse. Cell 85, 1009−1023.

Haldane, J.B.S., 1932. The Causes of Evolution. Longmans and Green, London.

Haldane, J.B.S., 1933. The part played by recurrent mutation in evolution. Am. Nat. 67, 5−9.

Hammoud, S.S., Nix, D.A., Zhang, H., et al., 2009. Distinctive chromatin in human sperm packages genes for embryo development. Nature 460, 473−478.

Han, M.V., Demuth, J.P., McGrath, C.L., et al., 2009. Adaptive evolution of young gene duplicates in mammals. Genome Res. 19, 859−867.

Hanahan, D., Weinberg, R.A., 2000. The hallmarks of cancer. Cell 100, 57−70.

Hanahan, D., Weinberg, R.A., 2011. Hallmarks of cancer: the next generation. Cell 144, 646—674.

Harach, H.R., Franssila, K.O., Wasenius, V.M., 1985. Occult papillary thyroid carcinoma of the thyroid: a "normal" finding in Finland. Cancer 56, 531—538.

Hardison, R.C., 1996. A brief history of hemoglobins: plant, animal, protist and bacteria. Proc. Natl. Acad. Sci. USA 93, 5675—5679.

Harris, H., 1985. Suppression of malignancy in hybrid cells: the mechanism. J. Cell Sci. 79, 83—94.

Harris, H., 2005. A long view of fashions in cancer research. BioEssays 27, 833—838.

Harris, H., Bramwell, M.E., 1987. The suppression of malignancy by terminal differentiation: evidence from hybrids between tumour cells and keratinocytes. J. Cell Sci. 87, 383—388.

Harris, J.R., 1991. The evolution of placental mammals. FEBS Lett. 295, 3—4.

Harshbarger, J.C., 1969. The registry of tumors in lower animals. Natl. Cancer Inst. Monogr. 31, XI—XVI.

Harshbarger, J.C., 1984. Pseudoneoplasms in ectothermic animals. Natl. Cancer Inst. Monogr. 65, 251—273.

Harshbarger, J.C., 1997. Invertebrate and cold-blooded vertebrate oncology. In: Rossi, L., Richardson, R., Harshberger., J. (Eds.), Spontaneous Animal Tumors: A Survey. Press Point di Abbiategrasso, Milano, pp. 41—53.

Hartl, M., Mitterstiller, A.-M., Valovka, T., et al., 2010. Stem cell-specific activation of an ancestral *myc* protooncogene with conserved basic functions in the early metazoan *Hydra*. Proc. Natl. Acad. Sci. USA 107, 4051—4056.

Hatsell, S., Frost, A.R., 2007. Hedgehog signaling in mammary gland development and breast cancer. J. Mammary Gland Biol. Neoplasia 12, 163—173.

Hayes, M.A., Ferguson, H.W., 1989. Neoplasia in fish. In: Ferguson, H.W. (Ed.), Systemic Pathology of Fish. Iowa State University Press, Ames (IA), pp. 230—247.

Hayflick, L., Moorhead, P.S., 1961. The serial cultivation of human diploid cell strains. Exp. Cell Res. 25, 585—621.

Haygood, R., Fedrigo, O., Hanson, B., et al., 2007. Promoter regions of many neural- and nutrition-related genes have experienced positive selection during human evolution. Nat. Genet. 39, 1140—1144.

He, X.L., Zheng, J.Z., 2005. Rapid subfunctionalization accompanied by prolonged and substantial neofunctionalization in duplicate gene evolution. Genetics 169, 1157—1164.

He, L.Z., Guidez, F., Tribioli, C., et al., 1998. Distinct interactions of PML-RARalpha and PLZF-RARalpha with co-repressors determine differential responses to RA in APL. Nat. Genet. 18, 126—135.

Heidebrecht, H.-J., Claviez, A., Kruse, M.L., et al., 2006. Characterization and expression of CT45 in Hodgkin's lymphoma (9 authors). Clin. Cancer Res. 12, 4804—4811.

Heidmann, O., Vernochet, C., Dupressoir, A., et al., 2009. Identification of an endogenous retroviral envelope gene with fusogenic activity and placenta-specific expression in rabbit: A new "syncytin" in a third order of mammals. Retrovirology 6, 107. Available from: http://dx.doi.org/10.1186/1742-4690-6-107.

Heinen, T.J.A.J., Staubach, F., Haming, D., et al., 2009. Emergence of a new gene from an intergenic region. Curr. Biol. 19, 1527—1531.

Hendrix, M.J.C., Seftor, E.A., Hess, A.R., et al., 2003. Vasculogenic mimicry and tumor-cell plasticity: lessons from melanoma. Nat. Rev. Cancer 3, 411—421.

Hendrix, M.J.C., Seftor, E.A., Seftor, R.E.B., et al., 2007. Reprogramming metastatic tumor cells with embryonic microenvironments. Nat. Rev. Cancer 7, 246—255.

Hengartner, M.O., Horvitz, H.R., 1994. *C. elegans* cell survival gene *ced-9* encodes a functional homolog of the mammalian proto-oncogene *bcl-2*. Cell 76, 665–676.

Hengartner, M.O., Ellis, R.E., Horvitz, H.R., 1992. *Caenorhabditis elegans* gene *ced-9* protects cells from programmed cell death. Nature 356, 494–499.

Hennen, G., 1965. Detection and study of a human-chorionic-thyroid-stimulating factor. Achs. Int. Physiol. Biochim. 73, 689–695.

Hennen, G., 1966a. The problem of thyroid-stimulating factors. Demonstration of a human placental thyrostimulin. Annls. Endocr. 27, 242–246.

Hennen, G., 1966b. Thyrotropin-like factor in a non-endocrine cancer tissue. Archs. Int. Physiol. Biochim 74, 701–704.

Hennen, G., 1966c. Detection of a thyroid-stimulating factor in choriocarcinoma occurring in the male. Archs. Int. Physiol. Biochim. 74, 303–307.

Herreros-Villanueva, M., Hijona, E., Cosme, A., et al., 2012. Spontaneous regression of pancreatic cancer: real or misdiagnosis? World J. Gastroenterol. 18, 2902–2908.

Hill, R.P., 1990. Tumor progression: potential role of unstable genomic changes. Cancer Metast. Rev. 9, 137–147.

Hlubek, F., Spaderna, S., Schmalhofer, O., et al., 2007. Wnt/FZD signaling and colorectal cancer morphogenesis. Front. Biosci. 12, 458–470.

Hoang, T., 2004. The origin of hematopoietic cell type diversity. Oncogene 23, 7188–7198.

Hobohm, U., 2001. Fever and cancer in perspective. Cancer Immunol. Immunother. 50, 391–396.

Hobohm, U., 2005. Fever therapy revisited. Br. J. Cancer 92, 421–425.

Hobohm, U., 2009. Healing heat: harnessing infection to fight cancer. Amer. Sci. 97, 34–41.

Hoekstra, H.E., 2012. Stickleback is the catch of the day. Nature 484, 46–47.

Hoekstra, H.E., Coyne, J.A., 2007. The locus of evolution: evo devo and the genetics of adaptation. Evolution 61, 995–1016.

Hoffman, F.M., Sternberg, P.W., Herskowitz, I., 1992. Learning about cancer genes through invertebrate genetics. Curr. Opin. Genet. Dev. 2, 45–52.

Hofmann, O., Caballero, O.L., Stevenson, B.J., et al., 2008. Genome-wide analysis of cancer-testis gene expression. Proc. Natl. Acad. Sci. USA 105, 20422–20427.

Holland, P.W.H., 1996. Molecular biology of lancelets: insight into development and evolution. Israel J. Zool 42, 247–272.

Holland, P.W.H., 2003. More genes in vertebrates? J. Struct. Funct. Genom. 3, 75–84.

Holland, P.W.H., Garcia-Fernandez, J., 1996. *Hox* genes and chordate evolution. Develop. Biol. 173, 382–395.

Holland, L.Z., Short, S., 2008. Gene duplication, co-option and recruitment during the origin of the vertebrate brain from the invertebrate chordate brain. Brain Behav. Evol. 72, 91–105.

Holt, A.B., Cheek, D.B., Mellits, E.D., et al., 1975. Brain size and the relation of the primate to the nonprimate. In: Cheek, D.B. (Ed.), Fetal and Postnatal Cellular Growth: Hormones And Nutrition. John Wiley, New York, pp. 23–44.

Holzschu, D., Lapierre, L.A., Lairmore, M.D., 2003. Comparative pathogenesis of epsilonretroviruses. Virology 77, 12385–12391.

Honjo, S., Doran, H.E., Stiller, C.A., et al., 2003. Neuroblastoma trends in Osaka, Japan, and Great Britain 1970–1994, in relation to screening. Int. J. Cancer 103, 538–543.

Hoover, K.L., 1984. Hyperplastic thyroid lesions in fish. Natl. Cancer Inst. Monogr. 65, 275–289.

Hosking, L., Tronsdale, J., Nicolai, H., et al., 1995. A somatic *BRCA1* mutation in an ovarian tumor. Nat. Genet. 9, 343–344.

Huang, M.E., Ye, Y.C., Chen, S.R., et al., 1987. All-trans retinoic acid with or without low dose cytosine arabinoside in acute promyelocytic leukemia. Report of 6 cases. Chin. Med. J. (Engl.) 100, 949–953.

Huang, M.E., Ye, Y.C., Chen, S.R., et al., 1988. Use of all-trans retinoic acid in the treatment of acute promyelocytic leukemia. Blood 72, 567–572.

Huang, H., Mahler-Araujo, B.M., Sankila, A., et al., 2000. APC mutations in sporadic medullo-blastomas. Am. J. Pathol. 156, 433–437.

Hubbard, S.A., Gargett, C.E., 2010. A cancer stem cell origin for human endometrial carcinoma? Reproduction 140, 23–32.

Huberman, E., Callaham, M.F., 1979. Induction of terminal differentiation in human promyelocytic leukemia cells by tumor-promoting agents. Proc. Natl. Acad. Sci. USA 76, 1293–1297.

Hughes, A.L., 1994. The evolution of functionally novel proteins after gene duplication. Proc. Biol. Sci. Lond. B. 256, 119–124.

Hughes, A.L., 2005. Gene duplication and the origin of novel proteins. Proc. Natl. Acad. Sci. USA 102, 8791–8792.

Hugo, H., Ackland, M.L., Blick, T., et al., 2007. Epithelial-mesenchymal and mesenchymal-epithelial transitions in carcinoma progression. J. Cell. Physiol. 213, 374–383.

Hurley, I., Hale, M.E., Prince, V.E., 2005. Duplication events and the evolution of segmental identity. Evol. Dev. 7, 556–567.

Hursting, S.D., Lavigne, J.A., Berrigan, D., et al., 2003. Calorie restriction, aging and cancer prevention: mechanisms of action and applicability to humans. Annu. Rev. Med. 54, 131–152.

Husseinzadeh, N., 2011. Status of tumor markers in epithelial ovarian cancer: has there been any progress? Gynecol. Oncol. 120, 152–157.

Huttley, G.A., Easteal, S., Southey, M.C., et al., Australian Breast Cancer Family Study 2000. Adaptive evolution of the tumor suppressor *BRCA1* in humans and chimpanzees. Nat. Genet. 25, 410–413.

Huxley, J., 1958. Biological Aspects of Cancer. George Allen & Unwin Ltd, London.

ICGSC, International Chicken Genome Sequence Consortium, 2004. Sequence and comparative analysis of the chicken genome provide unique perspectives on vertebrate evolution. Nature 432, 695–716.

IHGSC, International Human Genome Sequencing Consortium, 2004. Finishing the euchromatic sequence of the human genome. Nature 431, 931–945.

Ichikawa, Y., 1969. Differentiation of a cell line of myeloid leukemia. J. Cell. Physiol. 74, 223–234.

Innan, H., Kondrashov, F., 2010. The evolution of gene duplications: classifying and distinguishing between models. Nat. Rev. Genet 11, 97–108.

Irish, V.F., Litt, A., 2005. Flower development and evolution: gene duplication, diversification and redeployment. Curr. Opin. Genet. Dev. 15, 454–460.

Janic, A., Mendizabal, L., Llamazares, S., et al., 2010. Ectopic expression of germline genes drives malignant brain tumor growth in *Drosophila*. Science 330, 1824–1827.

Jeffery, C.J., 2009. Moonlighting proteins – an update. Mol. BioSyst. 5, 345–350.

Jenner, R.A., 2000. Evolution of animal body plans: the role of metazoan phylogeny at the interface between pattern and process. Evol. Dev. 2, 208–221.

Jensen, R.A., 1976. Enzyme recruitment in evolution of new function. Annu. Rev. Microbiol. 30, 409–425.

Jones, P.A., Baylin, S.B., 2002. The fundamental role of epigenetic events in cancer. Nat. Rev. Genet. 3, 415–428.

Jones, C.D., Begun, D.J., 2005. Parallel evolution of chimeric fusion genes. Proc. Natl. Acad. Sci. USA 102, 11373–11378.

Jones, P.A., Baylin, S.B., 2007. The epigenomics of cancer. Cell 128, 683–692.

Jones, F.C., Grabherr, M.G., Chan, Y.F., et al., 2012. The genomic basis of adaptive evolution in the threespine sticklebacks (29 authors). Nature 484, 55–61.

Jongeneel, C.V., Delorenzi, M., Iseli, C., et al., 2005. An atlas of human gene expression from massively parallel signature sequencing (MPSS). Genome Res. 15, 1007–1014.

Kaessmann, H., 2010. Origins, evolution, and phenotypic impact of new genes. Genome Res. 20, 1313–1326.

Kaessmann, H., Zollner, S., Nekrutenko, A., et al., 2002. Signatures of domain shuffling in the human genome. Genome Res. 12, 1642–1650.

Kaessmann, H., Vinckenbosch, N., Long, M., 2009. RNA-based gene duplication: mechanistic and evolutionary insights. Nat. Rev. 10, 19–31.

Kakizuka, A., Miller Jr., W.H., Umesono, K., et al., 1991. Chromosomal translocation t(15;17) in human acute promyelocytic leukemia fuses RAR alpha with a novel putative transcription factor, PML. Cell 66, 663–674.

Kalejs, M., Erenpreisa, J., 2005. Cancer/testis antigens and gametogenesis: a review and "brainstorming" session. Cancer Cell Int. 5, 4. Available from: http://dx.doi.org/10.1186/1475-2867-5-4.

Kapranov, P., Willingham, A.T., Gingeras, T.R., 2007. Genome-wide transcription and the implications for genomic organization. Nat. Rev. Genet. 8, 413–423.

Katayama, S., Tomaru, Y., Kasukawa, T., et al., 2005. Antisense transcription in the mammalian transcriptome. Science 309, 1564–1566.

Katsura, Y., Satta, Y., 2011. Evolutionary history of the cancer immunity antigen *MAGE* gene family. PLoS ONE 6 (6), e20365. Available from: http://dx.doi.org/10.1371/journal.pone.0020365.

Kavanagh, K.D., 2003. Perspective: embedded molecular switches, anticancer selection, and effects on ontogenetic rates: a hypothesis of developmental constraint on morphogenesis and evolution. Evolution 57, 939–948.

Kazazian Jr., H.H., 2004. Mobile elements: drivers of genome evolution. Science 303, 1626–1632.

Kazazian Jr., H.H., Wong, C., Youssoufian, H., et al., 1988. Haemophilia A resulting from *de novo* insertion of L1 sequences represents a novel mechanism for mutation in man. Nature 332, 164–166.

Kelley, M.L., Winge, P., Heaney, J.D., et al., 2001. Expression of homologues for p53 and p73 in the softshell clam (*Mya arenaria*), a naturally-occurring model for human cancer. Oncogene 20, 748–758.

Khalturin, K., Hemmrich, G., Fraune, S., et al., 2009. More than just orphans: are taxonomically-restricted genes important in evolution? Trends Genet. 25, 404–413.

Khaner, O., 2007. Evolutionary innovations of the vertebrates. Integr. Zool. 2, 60–67.

Kharin, A.V., Zagainova, I.V., Kostyuchenko, R.P., 2006. Formation of the zone of paratomy in freshwater oligochaets. Ontogenez 37, 424–437.

Khlopin, N.G., 1946. Biological and Experimental Bases of Histology. Publishing house of the Academy of Sciences of the U.S.S.R., Leningrad.

Kim, R., Emi, M., Tanabe, K., 2007. Cancer immunoediting from immune surveillance to immune escape. Immunology 121, 1–14.

Kimhi, Y., Palfrey, C., Spector, I., et al., 1976. Maturation of neuroblastoma cells in the presence of dimethylsulfoxide. Proc. Natl. Acad. Sci. USA 73, 462–466.

Kimura, M., 1980. A simple method for estimating evolutionary rates of base substitutions through comparative studies of nucleotide sequences. J. Mol. Evol. 16, 111–120.

King, M.C., Wilson, A.C., 1975. Evolution at two levels in humans and chimpanzees. Science 188, 107–116.

King, M.C., Marks, J.H., Mandel, J.B., The New York Breast Cancer Study Group, 2003. Breast and ovarian cancer risks due to inherited mutations in *BRCA1* and *BRCA2*. Science 302, 643–646.

Kingston, R.E., Baldwin, A.S., Sharp, P.A., 1985. Transcription control by oncogenes. Cell 41, 3–5.

Kinzler, K.W., Vogelstein, B., 1997. Gatekeepers and caretakers. Nature 386, 761–763.

Kisseljova, N.P., Kisseljov, F.L., 2005. DNA demethylation and carcinogenesis. Biochem. (Moscow) 70, 743–752.

Kleene, K.C., 2005. Sexual selection, genetic conflict, selfish genes, and the atypical patterns of gene expression in spermatogenic cells. Dev. Biol. 277, 16–26.

Kleene, K.C., Mulligan, E., Steiger, D., et al., 1998. The mouse gene encoding the testis-specific isoform of Poly(A) binding protein (Pabp2) is an expressed retroposon: intimations that gene expression in spermatogenic cells facilitates the creation of new genes. J. Mol. Evol. 47, 275–281.

Kluge, N., Ostertag, W., Sugiyama, T., et al., 1976. Dimethylsulfoxide-induced differentiation and hemoglobin synthesis in tissue cultures of rat erythroleukemia cells transformed by 7,12-dimethylbenz(a)anthracene. Proc. Natl. Acad. Sci. USA 73, 1237–1240.

Knoll, A.H., Carroll, S.B., 1999. Early animal evolution: emerging views from comparative biology and geology. Science 284, 2129–2137.

Knowles, D.G., McLysaght, A., 2009. Recent de novo origin of human protein-coding genes. Genome Res. 19, 1752–1759.

Knox, K., Baker, J.C., 2008. Genomic evolution of the placenta using co-option and duplication and divergence. Genome Res. 18, 695–705.

Knudson Jr., A.G., 1971. Mutation and cancer: statistical study of retinoblastoma. Proc. Natl. Acad. Sci. USA 68, 820–823.

Knudson Jr., A.G., 1979. Mutagenesis and embryonal carcinogenesis. Natl. Cancer Inst. Monogr. 51, 19–24.

Knudson Jr., A.G., 1989. Epidemiology of genetically determined cancer. In: Genetic Analysis of Tumour Suppression. Wiley, Chichester (Ciba Foundation Symposium 142), 3–19.

Knudson, A.G., 1996. Hereditary cancer: two hits revisited. J. Cancer Res. Clin. Oncol. 122, 135–140.

Koeffler, H.P., Bar-Eli, M., Territo, M., 1980. Phorbol diester-induced macrophage differentiation of leukemic blasts from patients with human myelogenous leukemia. J. Clin. Invest. 66, 1101–1108.

Korhonen, L., Brannvall, K., Skoglosa, Y., et al., 2003. Tumor suppressor gene *BRCA-1* is expressed by embryonic and adult neural stem cells and involved in cell proliferation. J. Neurosci. Res. 71, 769–776.

Korkaya, H., Paulson, A., Iovino, F., et al., 2008. HER2 regulates the mammary stem/progenitor cell population driving tumorigenesis and invasion. Oncogene 27, 6120–6130.

Kosswig, C., 1928. Uber bastarde der teleostier Platypoecilus und Xiphophorus. Z. Indukt. Abstammungs – Vererbungsl. 44, 150–158.

Kouprina, N., Pavlicek, A., Mochida, G.H., et al., 2004a. Accelerated evolution of the *ASPM* gene controlling brain size begins prior to human brain expansion. PLoS Biol. 2, 0653–0663.

Kouprina, N., Mullokandov, M., Rogozin, I.B., et al., 2004b. The *SPANX* gene family of cancer/testis-specific antigens: rapid evolution and amplification in African great apes and hominids. Proc. Natl. Acad. Sci. USA 101, 3077–3082.

Kouprina, N., Pavlicek, A., Collins, N.K., et al., 2005. The microcephaly *ASPM* gene is expressed in proliferating tissues and encodes for a mitotic spindle protein. Hum. Mol. Genet. 14, 2155–2165.

Kouprina, N., Noskov, V.N., Pavlicek, A., et al., 2007. Evolutionary diversification of SPANX-N sperm protein gene structure and expression (18 authors). PLoS ONE 2 (4), e359. Available from: http://dx.doi.org/10.1371/journal.pone.0000359.

Kowalski, P.E., Freeman, J.D., Nelson, D.T., et al., 1997. Genomic structure and evolution of a novel gene (PLA2L) with duplicated phospholipase A2-like domains. Genomics 39, 38–46.

Kozlov, A.P., 1976. Regulatory mechanisms as an expression and the result of evolution of competitive relations between the genes. In: Salinity Adaptations of the Aquatic Animals, 17 (25). Leningrad, Academy of Sciences of the U.S.S.R., p. 237.

Kozlov, A.P., 1979. Evolution of living organisms as a multilevel process. J. Theor. Biol 81, 1–17.

Kozlov, A.P., 1983. The principles of multi-level development of organisms. In: Maximov, V.N. (Ed.), Problems of Analysis of Biological Systems. Moscow University Press, Moscow.

Kozlov, A.P., 1987. Gene competition and the possible evolutionary role of tumors and cellular oncogenes. In: Presnov, Y.e.V., Maresin, Y.e.V., Zotin, A.I. (Eds.), Theoretical and Mathematical Aspects of Morphogenesis. Nauka, Moscow.

Kozlov, A.P., 1988. Conservation principles in the system of molecular-biological laws. Trans. Leningr. Soc. Nat. Sci. 87 (1), 4–21.

Kozlov, A.P., 1996. Gene competition and the possible evolutionary role of tumors. Med. Hypotheses 46, 81–84.

Kozlov, A.P., 2008. Tumors and evolution. Probl. Oncol. (Voprosy Onkologii) 54, 695–705.

Kozlov, A.P., 2010. The possible evolutionary role of tumors in the origin of new cell types. Med. Hypotheses 74, 177–185.

Kozlov, A., Krukovskaya, L., Baranova, A., et al., 2003. Transcriptional activation of evolutionary new genes in human tumors. Russ. J. HIV/AIDS and Related Problems 7 (1), 30–39, http://www.aidsconference.spb.ru/articles/arcB13808.pdf.

Kozlov, A.P., Emeljanov, A.V., Barabitskaja, O.V., et al., 1992. The maximal expression of mammalian genome, the complexity of tumor-specific transcripts and the cloning of tumor-specific cDNAs. Abstracts of annual meeting sponsored by laboratory of tumor cell biology. Bethesda.

Kozlov, A.P., Galachyants, Y.P., Dukhovlinov, I.V., et al., 2006. Evolutionarily new sequences expressed in tumors. Infect. Agent. Cancer 1, 8. Available from: http://dx.doi.org/10.1186/1750-9378-1-8.

Kozlov, A.P., Zabezhinski, M.A., Popovich, I.G., et al., 2012. Hyperplastic skin growth on the head of goldfish – comparative oncology aspects. Probl. Oncol. (Voprosi Oncologii) 58, 387–393.

Kramer, B.S., 2004. The science of early detection. Urol. Oncol. 22, 344–347.

Kramer, B.S., Croswell, J.M., 2009. Cancer screening: the clash of science and intuition. Annu. Rev. Med. 60, 125–137.

Krasnov, A.N., Kurshakova, M.M., Ramensky, V.E., et al., 2005. A retrocopy of a gene can functionally displace the source gene in evolution. Nucl. Acids Res. 33, 6654–6661. Available from: http://dx.doi.org/10.1093/nar/gki969.

Krebs Jr., E.T., 1946. Pathology of the trophoblast. J. Am. Med. Ass. 131, 1456.

Kreitman, M., 1996. The neutral theory is dead. Long live the neutral theory. BioEssays 18, 678–683.

Kreitman, M., Comeron, J.M., 1999. Coding sequence evolution. Curr. Opin. Genet. Dev. 9, 637–641.

Kriegstein, A., Noctor, S., Martinez-Cerdeno, V., 2006. Patterns of neural stem and progenitor cell division may underlie evolutionary cortical expansion. Nat. Rev. Neurosci. 7, 883–890.

Krukovskaya, L.L., Baranova, A., Tyezelova, T., et al., 2005. Experimental study of human expressed sequences newly identified in silico as tumor specific. Tumor Biol. 26, 17–24.

Krukovskaya, L.L., Nosova, Y.u.K., Polev, D.K., et al., 2007. Expression of nine tumor-associated nucleotide sequences in human normal and tumor tissues. Russ. J. AIDS, Cancer and Public Health 11, 117, http://www.aidsconference.spb.ru/articles/arcmtQbgT.pdf.

Krukovskaya, L.L., Shilov, E.S., Samusik, N.D., et al., 2009. Evolutionary new X1 and X2 genes are expressed in tumors. Russ. J. AIDS, Cancer and Public Health 13, 63, http://www.aidsconference.spb.ru/articles/arc9UuPPc.pdf.

Krukovskaya, L.L., Samusik, N.D., Shilov, E.S., et al., 2010. Tumor-specific expression of PBOV1, a new gene in evolution. Probl. Oncol. (Voprosy onkologii) 56, 327–332. Available from: http://www.ncbi.nlm.nih.gov/pubmed/20804056.

Krystosek, A., Sachs, L., 1976. Control of lysozyme induction in the differentiation of myeloid leukemic cells. Cell 9, 675–684.

Kryukov, G.V., Pennacchio, L.A., Sunyaev, S.R., 2007. Most rare missense alleles are deleterious in humans: implications for complex disease and association studies. Amer. J. Hum. Genet. 80, 727–739.

Kundu, S., Trent III, J.T., Hargrove, M.S., 2003. Plants, humans and hemoglobins. Trends Plant Sci. 8, 387–393.

Kurimoto, M., Fukuda, I., Hizuka, N., et al., 2008. The prevalence of benign and malignant tumors in patients with acromegaly at a single institute. Endocr. J. 55, 67–71.

Kusserow, A., Pang, K., Sturm, C., et al., 2005. Unexpected complexity of the *Wnt* gene family in a sea anemone. Nature 433, 156–160.

Kuwada, Y., 1911. Meiosis in the pollen mother cells of *Zea Mays*. L. Bot. Mag. 25, 163–181.

Lahn, B.T., Page, D.C., 1999. Four evolutionary strata on the human X chromosome. Science 286, 964–967.

Lala, P.K., Lee, B.P., Xu, G., et al., 2002. Human placental trophoblast as an in vitro model for tumor progression. Can. J. Physiol. Pharmacol. 80, 142–149.

Lander, E.S., Linton, L.M., Birren, B., et al., 2001. Initial sequencing and analysis of the human genome. Nature 409, 860–921.

Lapidot, T., Sirard, C., Vormoor, J., et al., 1994. A cell initiating human acute myeloid leukaemia after transplantation into SCID mice. Nature 367, 645–648.

Larsson, O., Girnita, A., Girnita, L., 2005. Role of insulin-like growth factor I receptor signaling in cancer. Brit. J. Cancer 92, 2097–2101.

Law, D.H., Liddle, G.W., Scott Jr., H.W., et al., 1965. Ectopic production of multiple hormones (ACTH, MSH and gastrin) by a single malignant tumor. New Engl. J. Med. 273, 292–296.

Leblanc, M.U., 1858. Recherches sur le cancer des animaux. Rec. Med. Vet. 35, 769–783.

Leder, A., Leder, P., 1975. Butyric acid, a potent inducer of erythroid differentiation in cultured erythroleukemic cells. Cell 5, 319–322.

Lee, P.N., Callaerts, P., de Couet, H.G., et al., 2003. Cephalopod *Hox* genes and the origin of morphological novelties. Nature 424, 1061–1065.

Lee, Y., McKinnon, P.J., 2000. ATM dependent apoptosis in the nervous system. Apoptosis 5, 523–529.

Lemons, D., McGinnis, W., 2006. Genomic evolution of Hox gene clusters. Science 313, 1918–1922.

Lengauer, C., Kinzler, K.W., Vogelstein, B., 1998. Genetic instabilities in human cancers. Nature 396, 643–649.

Lennon, G.G., Lehrach, H., 1991. Hybridization analyses of arrayed cDNA libraries. Trends Genet. 7, 314–317.

Leroi, A.M., Koufopanou, V., Burt, A., 2003. Cancer selection. Nat. Rev. Cancer 3, 226–231.

Leslie, M., 2011. Brothers in arms against cancer. Science 331, 1551–1552.

Leszczyniecka, M., Roberts, T., Dent, P., et al., 2001. Differentiation therapy of human cancer: basic science and clinical applications. Pharmacol. Ther. 90, 105–156.

Levine, M.T., Jones, C.D., Kern, A.D., et al., 2006. Novel genes derived from noncoding DNA in *Drosophila melanogaster* are frequently X-linked and exhibit testis-biased expression. Proc. Natl. Acad. Sci. USA 103, 9935–9939.

Lewis, B.P., Burge, C.B., Bartel, D.P., 2005. Conserved seed pairing, often flanked by adenosines, indicates that thousands of human genes are microRNA targets. Cell 120, 15–20.

Li, W.H., 1997. Molecular Evolution. Sinauer Associates, Sunderland, MA.

Li, J., Yen, C., Liaw, D., et al., 1997. *PTEN*, a putative protein tyrosine phosphatase gene mutated in human brain, breast, and prostate cancer. Science 275, 1943–1947.

Li, W.H., Gu, Z., Wang, H., et al., 2001. Evolutionary analyses of the human genome. Nature 409, 847–849.

Li, W.H., Yang, J., Gu, X., 2005. Expression divergence between duplicate genes. Trends Genet. 21, 602–607.

Li, L., Huang, Y., Xia, X., et al., 2006. Preferential duplication in the sparse part of yeast protein interaction network. Mol. Biol. Evol. 23, 2467–2473.

Li, C.Y., Zhang, Y., Wang, Z., et al., 2010. A human-specific de novo protein-coding gene associated with human brain functions. PLoS Comput. Biol. 6, e1000734. Available from: http://dx.doi.org/10.1371/journal.pcbi.1000734.

Liddle, G.W., Givens, J.R., Nicholson, W.E., et al., 1965. The ectopic ACTH syndrome. Cancer Res. 25, 1057–1061.

Lin, C.H., Hsieh, S.Y., Sheen, I.S., et al., 2001. Genome-wide hypomethylation in hepatocellular carcinogenesis. Cancer Res. 61, 4238–4243.

Lin, Q., Schwarz, J., Bucana, C., et al., 1997. Control of mouse cardiac morphogenesis and myogenesis by transcription factor MEF2C. Science 276, 1404–1407.

Lin, S.-Y., Elledge, S.J., 2003. Multiple tumor suppressor pathways negatively regulate telomerase. Cell 113, 881–889.

Lindeberg, S., 2009. Modern human physiology with respect to evolutionary adaptations that relate to diet in the past. In: Hublin, J.-J., Richards, M.P. (Eds.), The Evolution of Hominin Diets: Integrating Approaches to the Study of Palaeolithic Subsistence. Springer Science + Business Media B.V., pp. 43–57.

Lindesjoo, E., Thulin, J., 1994. Histopathology of skin and gills of fish in pulp mill effluents. Dis. Aquat. Org. 18, 81–93.

Lindeskog, M., Blomberg, J., 1997. Spliced human endogenous retroviral HERV-H env transcripts in T-cell leukaemia cell lines and normal leukocytes: alternative splicing pattern of HERV-H transcripts. J. Gen. Virol. 78 (Pt 10), 2575–2585.

Linnebacher, M., Maletzki, C., Klier, U., et al., 2012. Bacterial immunotherapy of gastrointestinal tumors. Langenbecks Arch. Surg. 397, 557–568.

Liotta, L., Petricoin, E., 2000. Molecular profiling of human cancer. Nat. Rev. Genet. 1, 48–56.

Liu, H.-K., Wang, Y., Belz, T., et al., 2010. The nuclear receptor tailless induces long-term neural stem cell expansion and brain tumor initiation. Genes Dev. 24, 683–695.

Liu, Y., Zhu, Q., Zhu, N., 2008. Recent duplication and positive selection of the GAGE gene family. Genetica 133, 31–35.

Livingstone, F.B., 1964. Aspects of the population dynamics of the abnormal hemoglobin and glucose-6-phosphate dehydrogenase deficiency genes. Am. J. Hum. Genet. 16, 435–450.

Lombard, L.S., Witte, E.J., 1959. Frequency and types of tumors in mammals and birds of the Philadelphia zoological garden. Cancer Res. 19, 127–141.

Long, M., 2007. Mystery genes. Nature 449, 511.

Long, M., Rosenberg, C., Gilbert, W., 1995. Intron phase correlation and the evolution of the intron/exon structure of genes. Proc. Natl. Acad. Sci. USA 92, 12495–12499.

Long, M., Betran, E., Thornton, K., et al., 2003. The origin of new genes: glimpses from the young and old. Nat. Rev. 4, 865–875.

Lotem, J., Sachs, L., 2002a. Cytokine control of developmental programs in normal hematopoiesis and leukemia. Oncogene 21, 3284–3294 (2002).

Lotem, J., Sachs, L., 2002b. Epigenetics wins over genetics: induction of differentiation in tumor cells. Semin. Cancer Biol. 12, 339–346.

Louicharoen, C., Patin, E., Paul, R., et al., 2009. Positively selected G6PD-Mahidol mutation reduces Plasmodium vivax density in Southeast Asians. Science 326, 1546–1549.

Lower, R., Lower, J., Tondera-Koch, C., et al., 1993. A general method for identification of transcribed retrovirus sequences (R-U5 PCR) reveals the expression of the human endogenous retrovirus loci HERV-H and HERV-K in teratocarcinoma cells. Virology 192, 501–511.

Lu, H., Zhu, G., Fan, L., et al., 2009. Etiology and pathology of epidermal papillomas in allogynogenetic crucian carp *Carassius auratus gibelio* (female) x *Cyprinus carpio var. singuonensis* (male). Dis. Aquat. Organ 83, 77–84.

Lui, J.H., Hansen, D.V., Kriegstein, A.R., 2011. Development and evolution of the human neocortex. Cell 146, 18–36.

Lund, A.H., van Lohuizen, M., 2002. RUNX: a trilogy of cancer genes. Cancer Cell 1, 213–215.

Luukko, K., Ylikorkala, A., Tiainen, M., et al., 1999. Expression of LKB1 and PTEN tumor suppressor genes during mouse embryonic development. Mech. Dev. 83, 187–190.

Lynch, M., 2007. The frailty of adaptive hypotheses for the origins of organismal complexity. Proc. Natl. Acad. Sci. USA 104, 8597–8604.

Lynch, M., Walsh, B., 1998. Genetics and Analysis of Quantitative Traits. Sinauer Associates, Sunderland, MA.

Lynch, M., Conery, J.S., 2000. The evolutionary fate and consequences of duplicate genes. Science 290, 1151–1155.

Lynch, M., Conery, J.S., 2003. The origins of genome complexity. Science 302, 1401–1404.

Lynch, M., O'Hely, M., Force, A., et al., 2001. The probability of preservation of newly arisen gene duplicate. Genetics 159, 1789–1804.

Machotka, S.V., McCain, B.B., Myers, M., 1989. Metastasis in fish. In: Kaiser, H.E. (Ed.), Comparative Aspects of Tumor Development. Kluwer, Dordrecht, pp. 48–54.

Maenhaut, C., Dumont, J.E., Roger, P.P., et al., 2010. Cancer stem cells: a reality, a myth, a fuzzy concept or a misnomer? An analysis. Carcinogenesis 31, 149–158.

Maher, B., 2012. The human encyclopaedia. Nature 489, 46–48.

Malakhov, V.V., 2004. Origin of bilateral-symmetrical animals (Bilateria). Zh. Obshch. Biol. 65, 371–388.

Mani, S.A., Guo, W., Liao, M.-J., et al., 2008. The epithelial-mesenchymal transition generates cells with properties of stem cells. Cell 133, 704–715.

Manne, U., Srivastava, R.G., Srivastava, S., et al., 2005. Recent advances in biomarkers for cancer diagnosis and treatment. DDT 10, 965–976.

Marchant, D.J., 2002. Benign breast disease. Obstet. Gynecol. Clin. North Am. 29, 1–20.

Markert, C.L., 1968. Neoplasia: a disease of cell differentiation. Cancer Res. 28, 1908–1914.

Marks, L.J., Russfield, A.B., Rosenbaum, D.L., 1963. Corticotropin in carcinoma of the lung. J. Am. Med. Ass. 183, 115–117.

Marotta, M., Piontkivska, H., Tanaka, H., 2012. Molecular trajectories leading to the alternative fates of duplicate genes. PLoS One 7, e38958.

Marques, A.C., Dupanloup, I., Vinckenbosch, N., et al., 2005. Emergence of young human genes after a burst of retroposition in primates. PLoS Biol. 3, 1970–1979.

Marques, A.C., Vinckenbosch, N., Brawand, D., et al., 2008. Functional diversification of duplicate genes through subcellular adaptation of encoded proteins. Genome Biol. 9, R54.

Marques-Bonet, T., Girirajan, S., Eichler, E.E., 2009a. The origins and impact of primate segmental duplications. Trends Genet. 25, 443–454.

Marques-Bonet, T., Ryder, O.A., Eichler, E.E., 2009b. Sequencing primate genomes: what have we learned? Annu. Rev. Genomics Hum. Genet. 10, 355–386.

Marsh, D.J., Dahia, P.L., Zheng, Z., et al., 1997. Germline mutations in *PTEN* are present in Bannayan-Zonana syndrome. Nat. Genet. 16, 333–334.

Martineau, D., Ferguson, H.W., 2006. Neoplasia. In: Ferguson, H.W. (Ed.), Systemic Pathology of Fish, 2nd edition.

Martinez-Morales, J.-R., Henrich, T., Ramialison, M., et al., 2007. New genes in the evolution of the neural crest differentiation program. Genome Biol. 8, R36. Available from: http://dx.doi.org/10.1186/gb-2007-8-3-r36.

Maurer, S., Koelmel, K.F., 1998. Spontaneous regression of advanced malignant melanoma. Onkologie 21, 14–18.

McCarrey, J.R., Thomas, K., 1987. Human testis-specific PGK gene lacks introns and possesses characteristics of a processed gene. Nature 326, 501–505.

McDonald, J.H., Kreitman, M., 1991. Adaptive protein evolution at the *Adh* locus in *Drosophila*. Nature 351, 652–654.

McLean, C.Y., Reno, P.L., Pollen, A.A., et al., 2011. Human-specific loss of regulatory DNA and the evolution of human-specific traits. Nature 471, 216–219.

McLysaght, A., Hokamp, K., Wolfe, K.H., 2002. Extensive genomic duplication during early chordate evolution. Nat. Genet. 31, 200–204.

McMahon, A.P., Bradley, A., 1990. The *Wnt-1* (*int-1*) proto-oncogene is required for development of a large region of the mouse brain. Cell 62, 1073–1085.

Mechler, B.M., 1994. In: Gordon, S. (Ed.), The Legacy of Cell Fusion. Oxford University Press, pp. 183–198, Ch. 13.

Mechler, B.M., McGinnes, W., Gehring, W.J., 1985. Molecular cloning of *lethal(2) giant larvae*: A recessive oncogene of *Drosophila melanogaster*. EMBO J. 4, 1551–1557.

Mechler, B.M., Torok, I., Schmidt, M., et al., 1989. Molecular basis for the regulation of cell fate by the *lethal(2) giant larvae* tumor suppressor gene of *Drosophila melanogaster*. Ciba Found. Symp. 142, 166–178.

Megason, S.G., McMahon, A.P., 2002. A mitogen gradient of dorsal midline Wnts organizes growth in the CNS. Development 129, 2087–2098.

Meizner, I., 2000. Perinatal oncology – the role of prenatal ultrasound diagnosis. Ultrasound Obstet. Gynecol. 16, 507–509.

Meizner, I., 2011. Introduction to fetal tumors. http://www.sonoworld.com/fetus/page.aspx? id = 515.

Mekel-Bobrov, N., Lahn, B.T., 2007. Response to comments by Timpson *et al.* and Yu *et al.* Science 317, 1036b.

Mekel-Bobrov, N., Gilbert, S.L., Evans, P.D., et al., 2005. Ongoing adaptive evolution of *ASPM*, a brain size determinant in *Homo sapiens*. Science 309, 1720–1722.

Mekel-Bobrov, N., Evans, P.D., Gilbert, S.L., et al., 2006. Response to comment on "Ongoing adaptive evolution of ASPM, a brain size determinant in *Homo sapiens*" and "*Microcephalin*, a gene regulating brain size, continues to evolve adaptively in humans." Science 313, 172b.

Meng, J., 2003. The journey from jaw to ear. Biologist 50, 54–158.

Meng, J., Wang, Y., Li, C., 2011. Transitional mammalian middle ear from a new Cretaceous Jehol eutriconodont. Nature 472, 181–185.

Merajver, S.D., Pham, T.M., Caduff, R.F., et al., 1995a. Somatic mutations in the *BRCA1* gene in sporadic ovarian tumors. Nat. Genet. 9, 439–443.

Merajver, S.D., Frank, T.S., Xu, J.Z., et al., 1995b. Germline *BRCA1* mutations and loss of the wild-type allele in tumors from families with early-onset breast and ovarian cancer. Clin. Cancer Res. 1, 539–544.

Merlo, L.M.F., Pepper, J.W., Reid, B.J., et al., 2006. Cancer as an evolutionary and ecological process. Nat. Rev. Cancer 6, 924–935.

Merrit, T.J.S., Quatro, J.M., 2001. Evidence for a period of directional selection following gene duplication in a neurally expressed locus of triosephosphate isomerase. Genetics 159, 689–697.

Messier, W., Stewart, C.B., 1997. Episodic adaptive evolution of primate lysozymes. Nature 385, 151–154.

Mi, S., Lee, X., Li, X., et al., 2000. Syncytin is a captive retroviral envelope protein involved in human placental morphogenesis. Nature 403, 785–789.

Michels, K.B., 2005. The role of nutrition in cancer development and prevention. Int. J. Cancer 114, 163–165.

Miki, Y., Swensen, J., Shattuckeidens, D., et al., 1994. A strong candidate for the breast and ovarian cancer susceptibility gene. Science 266, 66–71.

Mikkelsen, T.S., Hillier, L.W., Eichler, E.E., et al., 2005. Initial sequence of the chimpanzee genome and comparison with the human genome. Nature 437, 69–87.

Minguillon, C., Gardenyes, J., Serra, E., et al., 2005. No more than 14: the end of the amphioxus Hox cluster. Int. J. Biol. Sci. 1, 19–23.

Mintz, B., Illmensee, K., 1975. Normal genetically mosaic mice produced from malignant teratocarcinoma cells. Proc. Natl. Acad. Sci. USA 72, 3585–3589.

Mintz, B., Cronmiller, C., 1978. Normal blood cells of anemic genotype in teratocarcinoma-derived mosaic mice. Proc. Natl. Acad. Sci. USA 75, 6247–6251.

Miura, K., Sasaki, C., Katsushima, J., et al., 1966. Pituitary adrenal studies in patients with Cushing's syndrome induced by thymoma. J. Clin. Endocr. 27, 631–637.

Moczek, A.P., Rose, D., Sewell, W., et al., 2006. Conservation, innovation, and the evolution of horned beetle diversity. Dev. Genes Evol. 216, 655–665.

Modrek, B., Lee, C.J., 2003. Alternative splicing in the human, mouse and rat genomes is associated with an increased frequency of exon creation and/or loss. Nature Genet. 34, 177–180.

Molnar, Z., 2011. Evolution of cerebral cortical development. Brain Behav. Evol. 78, 94–107. Available from: http://dx.doi.org/10.1159/000327325.

Montgomery, S.H., Capellini, I., Venditti, C., et al., 2011. Adaptive evolution of four microcephaly genes and the evolution of brain size in anthropoid primates. Mol. Biol. Evol. 28, 625–638.

Moodie, R.L., 1917. Studies of paleopathology. I. General consideration of the evidences of pathological conditions found among fossil animals. Ann. Med. Hist. 1, 374–393.

Moore, R.C., Purugganan, M.D., 2003. The early stages of duplicate gene evolution. Proc. Natl. Acad. Sci. USA 100, 15682–15687.

Morales, P., Schmidt, R.E., 1991. Spindle-cell tumour resembling haemangiopericytoma in a common goldfish, *Carassius auratus* (L.) J. Fish Dis. 14, 499–502.

Moran, J.V., DeBerardinis, R.J., Kazazian Jr., H.H., et al., 1999. Exon shuffling by L1 retrotransposition. Science 283, 1530–1534.

Moran, N.A., Degnan, P.H., Santos, S.R., et al., 2005. The players in a mutualistic symbiosis: insects, bacteria, viruses, and virulence genes. Proc. Natl. Acad. Sci. USA 102, 16919–16926.

Morange, M., 2012. What is really new in the current evolutionary theory of cancer? J. Biosci. 37, 609–612.

Moreau-Aubry, A., Le Guiner, S., Labarriere, N., et al., 2000. A processed pseudogene codes for a new antigen recognized by a CD8(+) T cell clone on melanoma. J. Exp. Med. 191, 1617–1624.

Morel, A.-P., Lievre, M., Thomas, C., et al., 2008. Generation of breast cancer stem cells through epithelial-mesenchymal transition. PLoS ONE 3, e2888.

Morris, H.P., Wagner, B.P., 1968. Induction and transplantation of rat hepatomas with different growth rate (including minimal deviation hepatomas). Methods in Cancer Res. IV, 125–152.

Morris, H.P., Meranze, D.R., 1974. Induction and some characteristics of "minimal deviation" and other transplantable rat hepatomas. Recent Results Cancer Res. 44, 103–111.

Morrison, S.J., Spradling, A.C., 2008. Stem cells and niches: mechanisms that promote stem cell maintenance throughout life. Cell 132, 598–611.

Muller, J., 1838. Ueber den feinern Bau und die Formen der Krankhaften Geschwulste. G. Reimer, Berlin.

Muller, H.J., 1935. The origin of chromatin deficiencies as minute deletions subject to insertion elsewhere. Genetics 17, 237–252.

Muller, G.B., 2007. Evo-devo: extending the evolutionary synthesis. Nat. Rev. Genet. 8, 943–949.

Muller, R., Verma, I.M., 1984. Expression of cellular oncogenes. Curr. Top. Microbiol. Immunol. 112, 73–115.

Muller, G.B., Wagner, G.P., 1991. Novelty in evolution: restructuring the concept. Annu. Rev. Ecol. Syst. 22, 229–256.

Muller, G.B., Newman, S.A., 2005. Evolutionary innovation and morphological novelty. J. Exp. Zool. (Mol. Dev. Evol.) 304B, 485–486.

Muotri, A.R., Gage, F.H., 2006. Generation of neuronal variability and complexity. Nature 441, 1087–1093.

Muotri, A.R., Chu, V.T., Marchetto, M.C.N., et al., 2005. Somatic mosaicism in neuronal precursor cells mediated by L1 retrotransposition. Nature 435, 903–910.

Muttray, A.F., Cox, R.L., St-Jean, S.D., 2005. Identification and phylogenetic comparison of p53 in two distinct mussel species (*Mytilus*). Comp. Biochem. Physiol. C Pharmacol. Toxicol. Endocrinol. 140, 237–250.

Muttray, A.F., Cox, R.L., Reinisch, C.L., et al., 2007. Identification of DeltaN isoform and polyadenylation site choice variants in molluscan p63/p73-like homologues. Mar. Biotechnol. 9, 217–230.

Muttray, A.F., Schulte, P.M., Baldwin, S.A., et al., 2008. Invertebrate p53-like mRNA isoforms are differentially expressed in mussel haemic neoplasia. Mar. Environ. Res. 66, 412−421.

Muttray, A.F., O'Toole, T.F., Morrill, W., et al., 2010. An invertebrate *mdm* homolog interacts with p53 and is differentially expressed together with *p53* and *ras* in neoplastic *Mytilus trossulus* haemocytes. Comp. Biochem. Physiol., Part B 156, 298−308.

Naguib, F.N., el Kouni, M.H., Cha, S., 1985. Enzymes of uracil catabolism in normal and neoplastic human tissues. Cancer Res. 45, 5405−5412.

Narod, S.A., Ford, D., Devilee, P., et al., 1995. An evaluation of genetic heterogeneity in 145 breast-ovarian cancer families. Am. J. Hum. Genet. 56, 254−264.

Nauts, H.C., Fowler, G.A., Bogatko, F.H., 1953. A review of the influence of bacterial infection and of bacterial products (Coley's toxins) on malignant tumors in man. Acta Med. Scand. 145 (Suppl. 276), 1−103.

Nekrutenko, A., 2004. Identification of novel exons from rat-mouse comparisons. J. Mol. Evol. 59, 703−708.

Neugut, A.I., Jacobson, J.S., Rella, V.A., 1997. Prevalence and incidence of colorectal adenomas and cancer in asymptomatic persons. Gastrointest. Endosc. Clin. N. Am. 7, 387−399.

Nichols, J., Warren, J.C., Adams, F.A., 1962. ACTH-like excretion from carcinoma of the ovary. J. Am. Med. Ass. 182, 713−718.

Nielsen, C., 1995. Animal Evolution: Interrelationships of the Living Phyla. Oxford University Press, Oxford.

Nielsen, C., Norrevang, A., 1985. The trochaea theory: an example of life cycle phylogeny. In: Conway Morris, S., George, J.D., Gibson, R., Platt, H.M. (Eds.), The Origin and Relationships of Lower Invertebrate Groups. Oxford University Press, Oxford, pp. 28−41.

Nielsen, R., Bustamante, C., Clark, A.G., et al., 2005. A scan for positively selected genes in the genomes of humans and chimpanzees. PLoS Biol. 6, e170.

Nielsen, R., Hellmann, I., Hubisz, M., et al., 2007. Recent and ongoing selection in the human genome. Nat. Rev. Genet. 8, 857−868.

Nilsson, B., 1984. Probable in vivo induction of differentiation by retinoic acid of promyelocytes in acute promyelocytic leukaemia. Br. J. Haematol. 57, 365−371.

Noctror, S.C., Martinez-Cerdeno, V., Ivic, L., et al., 2004. Cortical neurons arise in symmetric and asymmetric division zones and migrate through specific phases. Nat. Neurosci. 7, 136−144.

Noga, E.J., 2000. Fish Disease. Diagnosis and Treatment. Wiley-Blackwell, Hoboken (NJ), 367p.

Nowell, P.C., 1976. The clonal evolution of tumor cell populations. Science 194, 23−28.

Nowell, P.C., 1986. Mechanisms of tumor progression. Cancer Res. 46, 2203−2207.

Nusse, R., 2001. The *Wnt* gene homepage. http://www-leland.stanford.edu/ ∼ rnusse/wtnwindow.html.

Nusse, R., Varmus, H.E., 1982. Many tumors induced by mouse mammary tumor virus contain a provirus integrated in the same region of the host chromosome. Cell 31, 99−109.

Nusse, R., Varmus, H.E., 1992. *Wnt* genes. Cell 69, 1073−1087.

Oberling, C., 1946. The Riddle of Cancer. Yale Univ. Press, New Haven.

Odell, W.D., Hertz, R., Lipsett, M.B., 1967. Endocrine aspects of trophoblastic neoplasms. Clin. Obstet. Gynec. 10, 290−302.

Ohno, S., 1970. Evolution by Gene Duplication. Springer-Verlag, New York, Heidelberg, Berlin, 160 pp.

Ohno, S., 1973. Ancient linkage groups and frozen accidents. Nature 244, 259−262.

Ohta, T., 1991. Multigene families and the evolution of complexity. J. Mol. Evol. 33, 34−41.

Ohta, T., 1992. The nearly neutral theory of molecular evolution. Ann. Rev. Ecol. Syst. 23, 263–286.

Olson, E.N., 2006. Gene regulatory networks in the evolution and development of the heart. Science 313, 1922–1927.

Omenn, G.S., 1970. Ectopic polypeptide hormone production by tumors. Ann. Intern. Med. 72, 136–138.

Orians, G.H., 1980. Habitat selection: general theory and application to human behavior. In: Lockard, J.S. (Ed.), The Evolution of Human Social Behavior. Elsevier Press, New York, pp. 49–66.

Pal, C., Papp, B., Hurst, L.D., 2001. Highly expressed genes in yeast evolve slowly. Genetics 158, 927–931.

Palena, C., Polev, D.E., Tsang, K.Y., et al., 2007. The human T-box mesodermal transcription factor Brachyury is a candidate target for T-cell-mediated cancer immunotherapy. Clin. Cancer Res. 13, 2471–2478.

Pandha, H.S., Waxman, J., 1995. Tumor markers. Q. J. Med. 88, 233–241.

Pani, A.M., Mullarkey, E.E., Aronowicz, J., et al., 2012. Ancient deuterostome origins of vertebrate brain signaling centers. Nature 483, 289–294.

Papaioannou, V.E., McBurney, M.W., Gardner, R.L., et al., 1975. Fate of teratocarcinoma cells injected into early mouse embryos. Nature 258, 70–73.

Pardal, R., Molofsky, A.V., He, S., et al., 2005. Stem cell self-renewal and cancer cell proliferation are regulated by common networks that balance the activation of proto-oncogenes and tumor suppressors. Cold Spring Harb. Symp. Quant. Biol. 70, 177–185.

Parr, B.A., Shea, M.J., Vassileva, G., et al., 1993. Mouse Wnt genes exhibit discrete domains of expression in the early embryonic and limb buds. Development 119, 247–261.

Parra, G., Reymond, A., Dabbouseh, N., et al., 2006. Tandem chimerism as a means to increase protein complexity in the human genome. Genome Res. 16, 37–44.

Parsons, D.W., Li, M., Zhang, X., et al., 2011. The genetic landscape of the childhood cancer medulloblastoma. Science 331, 435–439.

Pathak, S., Multani, A.S., 2006. Aneuploidy, stem cells and cancer. EXS 2006, 49–64.

Patthy, L., 1985. Evolution of the proteases of blood coagulation and fibrinolysis by assembly from modules. Cell 41, 657–663.

Patthy, L., 1996. Exon shuffling and other ways of module exchange. Matrix Biol. 15, 301–310, 311–312.

Patthy, L., 2003. Modular assembly of genes and the evolution of new functions. Genetica 118, 217–231.

Paulding, C.A., Ruvolo, M., Haber, D.A., 2003. The Tre2 (USP6) oncogene is a hominoid-specific gene. Proc. Natl. Acad. Sci. USA 100, 2507–2511.

Pavlicek, A., Noskov, V., Kouprina, N., et al., 2004. Evolution of the tumor suppressor BRCA1 locus in primates: implications for cancer predisposition. Hum. Mol. Genet. 13, 2737–2751.

Peifer, M., Polakis, P., 2000. Wnt signaling in oncogenesis and embryogenesis – a look outside the nucleus. Science 287, 1606–1609.

Perkins, A., Kongsuwan, K., Visvader, J., et al., 1990. Homeobox gene expression plus autocrine growth factor production elicits myeloid leukemia. Proc. Natl. Acad. Sci. USA 87, 8398–8402.

Peterson, K.J., Davidson, E.H., 2000. Regulatory evolution and the origin of the bilaterians. Proc. Natl. Acad. Sci. USA 97, 4430–4433.

Petersen, C.P., Reddien, P.W., 2011. Polarized notum activation at wounds inhibits Wnt function to promote planarian head regeneration. Science 332, 852–855.

Peterson, K.J., Irvine, S.Q., Cameron, R.A., et al., 2000a. Quantitative assessment of *Hox* complex expression in the indirect development of the polychaete annelid *Chaetopterus sp.* Proc. Natl. Acad. Sci. USA 97, 4487−4492.

Peterson, K.J., Cameron, R.A., Davidson, E.H., 2000b. Bilaterian origins: significance of new experimental observations. Dev. Biol. 219, 1−17.

Peterson, K.J., McPeek, M.A., Evans, D.A.D., 2005. Tempo and mode of early animal evolution: inferences from rocks, Hox, and molecular clocks. Paleobiology 31 (2, Supplement), 36−55.

Peverali, F.A., D'Esposito, M., Acampora, D., et al., 1990. Expression of HOX homeogenes in human neuroblastoma cell culture lines. Differentiation 45, 61−69.

Pfeiffer, C.J., Nagai, T., Fujimura, M., et al., 1979. Spontaneous regressive epitheliomas in the Japanese newt, *Cynops pyrrhogaster*. Cancer Res. 39, 1904−1910.

Piatigorsky, J., Wistow, G., 1991. The recruitment of crystallins: new functions precede gene duplication. Science 252, 1078−1079.

Pickrell, J.K., Coop, G., Novembre, J., et al., 2009. Signals of recent positive selection in a worldwide sample of human populations (11 authors). Genome Res. 19, 826−837.

Pierce, G.B., 1983. The cancer cell and its control by the embryo. Am. J. Pathol. 113, 117−124.

Pierce, G.B., 1993. In: Arechaga, J. On the boundary between development and neoplasia. An interview with Professor G. Barry Pierce. Int. J. Dev. Biol. 37, 5−16.

Pierce, G.B., Dixon, F.J., 1959. Testicular teratomas. I. Demonstration of teratogenesis by metamorphosis of multipotential cells. Cancer 12, 573−583.

Pierce Jr., G.B., Verney, E.L., 1961. An in vitro and in vivo study of differentiation in teratocarcinoma. Cancer 14, 1017−1029.

Pierce Jr., G.B., Dixon Jr., F.J., Verney, E.L., 1960. Teratocarcinogenic and tissue-forming potentials of cell types comprising neoplastic embryoid bodies. Lab. Invest. 9, 583−602.

Pierce, G.B., Nakane, P.K., Martinez-Hernandez, A., et al., 1977. Ultrastructural comparison of differentiation of stem cells of murine adenocarcinomas of colon and breast with their normal counterparts. J. Natl. Cancer Inst. 58, 1329−1345.

Pierce, G.B., Shikes, R., Fink, L.M., 1978. Cancer: A Problem of Developmental Biology. Prentice-Hall, New York.

Pierce, G.B., Pantazis, C.G., Caldwell, J.E., et al., 1982. Specificity of the control of tumor formation by the blastocyst. Cancer Res. 42, 1082−1087.

Piper, W., 1960. Cushing's syndrome in primary ectopic chorioepithelioma of the liver. Internist 1, 420−424.

Podesta, A.H., Mullins, J., Pierce, G.B., et al., 1984. The neural stage mouse embryo in control of neuroblastoma. Proc. Natl. Acad. Sci. USA 81, 7608−7611.

Pogribny, I.P., Beland, F.A., 2009. DNA hypomethylation in the origin and pathogenesis of human diseases. Cell. Mol. Life Sci. Available from: http://dx.doi.org/10.1007/s00018-009-0015-5.

Polakis, P., 2000. Wnt signaling and cancer. Genes & Dev. 14, 1837−1851.

Polev, D.E., Nosova, J.K., Krukovskaya, L.L., et al., 2009. Expression of transcripts corresponding to cluster Hs.633957 in human healthy and tumor tissues. Mol. Biol. (Mosk) 43, 88−92.

Polev, D.E., Krukovskaya, L.L., Kozlov, A.P., 2011a. Expression of the locus Hs.633957 in human digestive system and tumors. Probl. Oncol. (Voprosy Onkologii) 57, 48−49.

Polev, D., Krukovskaia, L., Karnaukhova, J., et al., 2011b. Transcribed locus Hs.633957: A new tumor-associated primate-specific gene with possible microRNA function. Proceedings of the 102nd Annual Meeting of the American Association for Cancer Research. Abstract No. 3858.

Polev, D.E., Karnaukhova, J.K., Krukovskaya, L.L., et al., 2014. *ELFN1-AS1* — a novel primate gene with possible microRNA function expressed predominantly in tumors. BioMed Res. Internatl., in press.

Pomiankowski, A., 1999. Intragenomic conflict. In: Keller, L. (Ed.), Levels of Selection in Evolution. Princeton University Press, Princeton.

Ponting, C., Jackson, A.P., 2005. Evolution of primary microcephaly genes and the enlargement of primate brains. Curr. Opin. Genet. & Dev. 15, 241–248.

Ponting, C.P., Oliver, P.L., Reik, W., 2009. Evolution and functions of long noncoding RNAs. Cell 136, 629–641.

Popp, C., Dean, W., Feng, S., et al., 2010. Genome-wide erasure of DNA methylation in mouse primordial germ cells is affected by AID deficiency. Nature 463, 1101–1105.

Potter, V.R., 1978. Phenotypic diversity in experimental hepatomas: the concept of partially blocked ontogeny. Br. J. Cancer 38, 1–23.

Potter, V.R., 1981. The present status of the blocked ontogeny hypothesis of neoplasia: the thalassemia connection. Oncodevel. Biol. Med. 2, 243–266.

Poulet, F.M., Wolfe, M.J., Splitsbergen, J.M., 1994. Naturally occurring orocutaneous papillomas and carcinomas of brown bullheads (*Ictalurus nebulosis*) in New York state. Vet. Pathol. 31, 8–18.

Preisler, H.D., Lyman, G., 1975. Differentiation of erythroleukemia cells in vitro: properties of chemical inducers. Cell Differ. 4, 179–185.

Prensner, J.R., Chinnaiyan, A.M., 2011. The emergence of lncRNAs in cancer biology. Cancer Disc. 1, 391–407.

Priester, W.A., 1980. Sources of data. Natl. Cancer Inst. Monogr. 54, 1–10.

Priester, W.A., McKay, F.W., 1980. The occurrence of tumors in domestic animals. Natl. Cancer Inst. Monogr. 54, 1–210.

Prince, V.E., Pickett, F.B., 2002. Splitting pairs: the diverging fates of duplicated genes. Nat. Rev. Genet. 3, 827–837.

Prud'homme, B., Gompel, N., Carroll, S.B., 2007. Emerging principles of regulatory evolution. Proc. Natl. Acad. Sci. USA 104, 8605–8612.

Prud'homme, B., Minervino, C., Hocine, M., et al., 2011. Body plan innovation in treehoppers through the evolution of an extra wing-like appendage. Nature 473, 83–86.

Puente, X.S., Pinyol, M., Quesada, V., et al., 2011. Whole-genome sequencing identifies recurrent mutations in chronic lymphocytic leukaemia. Nature 475, 101–105.

Quint, M., Drost, H.-G., Gabel, A., et al., 2012. A transcriptomic hourglass in plant embryogenesis. Nature 490, 98–101.

Rabbitts, T.H., 1991. Translocations, master genes, and differences between the origins of acute and chronic leukemias. Cell 67, 641–644.

Rabinovsky, R., Uhr, J.W., Vitetta, E.S., et al., 2007. Cancer dormancy: lessons from a B cell lymphoma and adenocarcinoma of the prostate. Adv. Cancer Res. 97, 189–202.

Raff, R.A., 1996. The Shape of Life. The University of Chicago Press, Chicago.

Raff, R.A., 2008. Origins of the other metazoan body plans: The Evolution of Larval Forms. Phil. Trans. R. Soc. B 363, 1473–1479.

Rajewsky, N., 2006. microRNA target predictions in animals. Nat. Genet. 38 (Suppl 1), S8–13.

Rakic, P., 1995. A small step for the cell, a giant leap for mankind: a hypothesis of neocortical expansion during evolution. Trends Neurosci. 18, 383–388.

Rassoulzadegan, M., Grandjean, V., Gounon, P., et al., 2006. RNA-mediated non-mendelian inheritance of an epigenetic change in the mouse. Nature 441, 469–474.

Rastogi, S., Liberles, D.A., 2005. Subfunctionalization of duplicated genes as a transition state to neofunctionalization. BMC Evol. Biol. 5, 28.

Ratcliffe, H.L., 1933. Incidence and nature of tumors in captive wild animals and birds. Amer. J. Cancer 17, 116–135.

Rausch, R.L., Rausch, V.R., 1968. On the biology and systematic position of *Microtus abbreviatus* Miller, a vole endemic to the St. Matthew islands, Bering sea. Z. Saugetierkunde 33, 65–99.

Reanney, D.C., 1974. Viruses and evolution. Int. J. Cytol. 37, 21–52.

Rebeiz, M., Jikomes, N., Kassner, V.A., et al., 2011. Evolutionary origin of a novel gene expression pattern through co-option of the latent activities of existing regulatory sequences. Proc. Natl. Acad. Sci. USA 108, 10036–10043.

Rehen, S.K., McConnell, M.J., Kaushal, D., et al., 2001. Chromosomal variation in neurons of the developing and adult mammalian nervous system. Proc. Natl. Acad. Sci. USA 98, 13361–13366.

Reiss, M., Gamba-Vitalo, C., Sartorelli, A.C., et al., 1986. Induction of tumor cell differentiation as a therapeutic approach: preclinical models for hematopoietic and solid neoplasms. Cancer Treat. Rep. 70, 201–218.

Renan, M.J., 1993. How many mutations are required for tumorigenesis? Implications from human cancer data. Mol. Carcinog. 7, 139–146.

Reuben, R.C., Wife, R.L., Breslow, R., et al., 1976. A new group of potent inducers of differentiation in murine erythroleukemia cells. Proc. Natl. Acad. Sci. USA 73, 862–866.

Reya, T., Clevers, H., 2005. Wnt signaling in stem cells and cancer. Nature 434, 843–850.

Reya, T., Morrison, S.J., Clarke, M.F., et al., 2001. Stem cells, cancer, and cancer stem cells. Nature 414, 105–111.

Reynolds, L.P., Borowicz, P.P., Vonnahmae, K.A., et al., 2005. Animal models of placental angiogenesis. Placenta 26, 689–708.

Rhodes, D.R., Yu, J., Shanker, K., et al., 2004. Large scale meta-analysis of cancer microarray data identifies common transcriptional profiles of neoplastic transformation and progression. Proc. Natl. Acad. Sci. USA 101, 9309–9314.

Roberts, R.J., 2001. Neoplasia of teleosts. In: Roberts, R.J. (Ed.), Fish Pathology, third ed. W.B. Saunders, London, pp. 151–168.

Robertson, K.D., 2001. DNA methylation, methyltransferases and cancer. Oncogene 20, 3139–3155.

Rodriguez-Trellers, F., Tarrio, R., Ayala, F.J., 2003. Convergent neofunctionalization by positive Darwinian selection after ancient recurrent duplications of the xanthine *dehydrogenase* gene. Proc. Natl. Acad. Sci. USA 100, 13413–13417.

Rohwer, F., Thurber, R.V., 2009. Viruses manipulate the marine environment. Nature 459, 207–212.

Rolig, R.L., McKinnon, P.J., 2000. Linking DNA damage and neurodegeneration. Trends Neurosci. 23, 417–424.

Rosen, S.W., Becker, C.F., Schlaff, S., et al., 1968. Ectopic gonadotropin production before clinical recognition of bronchogenic carcinoma. New Engl. J. Med. 279, 640–641.

Rosenblueth, A., Wiener, N., Bigelow, J., 1943. Behavior, purpose and teleology. Philos. Sci. 10, 18–24.

Ross, M.T., Grafham, D.V., Coffey, A.J., et al., 2005. The DNA sequence of the human X chromosome. Nature 434, 325–337.

Rosso, L., Margues, A.C., Weier, M., et al., 2008a. Birth and rapid subcellular adaptation of hominoid-specific CDC14 protein. PLoS Biol. 6, e140.

Rosso, L., Marques, A.C., Reichert, A.S., et al., 2008b. Mitochondrial targeting adaptation of the hominoid-specific glutamate dehydrogenase driven by positive Darwinian selection. PLoS Genet. 4, e1000150.

Rothschild, B.M., Tanke, D.H., Helbling, M., et al., 2003. Epidemiologic study of tumors in dinosaurs. Naturwissenschaften 90, 495–500.

Rous, P., 1910. A transmissible avian neoplasm (sarcoma of the common fowl). J. Exp. Med. 12, 696–705.

Rous, P., 1911. A sarcoma of the fowl transmissible by an agent separable from the tumor cells. J. Exp. Med. 13, 347–411.

Rous, P., Beard, J.W., 1935. The progression to carcinoma of virus-induced rabbit papillomas (Shope). J. Exp. Med. 62, 523–548.

Rous, P., Murphy, J.B., 1913. Variation in a chicken sarcoma caused by a filterable agent. J. Exp. Med. 17, 219–231.

Rowley, J.D., Golomb, H.M., Dougherty, C., 1977. 15/17 translocation, a consistent chromosomal change in acute promyelocytic leukaemia. Lancet 1, 549–550.

Roy, N.K., Kreamer, G.L., Konkle, B., et al., 1995. Characterization and prevalence of a polymorphism in the 3' untranslated region of cytochrome P4501A1 in cancer-prone Atlantic tomcod. Arch. Biochem. Biophys. 322, 204–213.

Ruberte, E., Dolle, P., Krust, A., et al., 1990. Specific spacial and temporal distribution of retinoic acid receptor gamma transcripts during mouse embryogenesis. Development 108, 213–222.

Ruberte, E., Dolle, P., Chambon, P., et al., 1991. Retinoic acid receptors and cellular retinoid binding proteins. II. Their different pattern of transcription during early morphogenesis in mouse embryos. Development 111, 45–60.

Rubin, I.C., 1910. Pathological diagnosis of incipient carcinoma of uterus. Am. J. Obst. 62, 668–676.

Rubin, G.M., Yandell, M.D., Wortman, J.R., et al., 2000. Comparative genomics of the Eukaryotes (54 authors). Science 287, 2204–2215.

Rubin, I.C., 1918. Pathogenesis and further growth of carcinoma of uterus in relation to clinical symptoms and early diagnosis. Am. J. Obst. 78, 353–373.

Ruiz, I., Altaba, A., Jessel, T., 1991. Retinoic acid modifies mesodermal patterning in early Xenopus embryos. Genes Dev. 5, 175–187.

Russo, P., Uzzo, R.G., Lowrance, W.T., et al., 2012. Incidence of benign versus malignant renal tumors in selected studies. J. Clin. Oncol. 30 (suppl. 5), 357.

Rutherford, S.L., Lindquist, S., 1998. Hsp90 as a capacitor for morphological evolution. Nature 396, 336–342.

Ryan, P.D., Goss, P.E., 2008. The emerging role of the insulin-like growth factor pathway as a therapeutic target in cancer. Oncologist 13, 16–24.

Ryan, J.F., Mazza, M.E., Pang, K., et al., 2007. Pre-bilaterian origins of the Hox cluster and the Hox code: evidence from the sea anemone, Nematostella vectensis. PLoS ONE 1, e153.

Ryder, J.A., 1887. On a tumor in the oyster. Proc. Natl. Acad. Sci. USA 44, 25–27.

Sabeti, P.C., Schaffner, S.F., Fry, B., et al., 2006. Positive natural selection in the human lineage. Science 312, 1614–1620.

Sabeti, P.C., Varilly, P., Fry, B., Lohmueller, J., Hostetter, E., Cotsapas, C., and The International HapMap Consortium, 2007. Genome-wide detection and characterization of positive selection in human populations. Nature 449, 913–918.

Sachs, L., 1978. Control of normal cell differentiation and the phenotypic reversion of malignancy in myeloid leukemia. Nature 274, 535–539.

Sachs, L., 1996. The control of hematopoiesis and leukemia: from basic biology to the clinic. Proc. Natl. Acad. Sci. USA 93, 4742–4749.

Sahin, U., Tureci, O., Schmitt, H., et al., 1995. Human neoplasms elicit specific immune responses in the autologous host. Proc. Natl. Acad. Sci. USA 92, 11810–11813.

Sahin, U., Tureci, O., Chen, Y.T., et al., 1998. Expression of multiple cancer/testis antigens in breast cancer and melanoma: basis for polyvalent CT vaccine strategies. Int. J. Cancer 78, 387–389.

Salim, O.E., Hussein, S., Ibrahim, S.Z., et al., 2009. Spontaneous regression of liver metastasis. Khartoum Med. J. 2, 178–180.

Samusik, N.A., Galachyantz, Y.P., Kozlov, A.P., 2007. Comparative-genomic analysis of human tumor-related transcripts. Russ. J. AIDS, Cancer and Public Health 10 (2), 61–62, http://www.aidsconference.spb.ru/articles/arcmtQbgT.pdf.

Samusik, N.A., Galachyants, Y.P., Kozlov, A.P., 2009. Analysis of evolutionary novelty of tumor-specifically expressed sequences. Ecologicheskaya Genetika 7, 26–37.

Samusik, N.A., Galachyants, Y.P., Kozlov, A.P., 2011. Analysis of evolutionary novelty of tumor-specifically expressed sequences. Russ. J. Genet.: Applied Res. 1, 138–148.

Samusik, N., Krukovskaya, L., Meln, I., et al., 2013. PBOV1 is a human de novo gene with tumor-specific expression that is associated with a positive clinical outcome of cancer. PLOS ONE 8, e56162.

Sanchez Alvarado, A., Newmark, P.A., 1998. The use of planarians to dissect the molecular basis of metazoan regeneration. Wound Repair Regen. 6, 413–420.

Santisteban, M., Reiman, J.M., Asiedu, M.K., et al., 2009. Immune-induced epithelial to mesenchymal transition in vivo generates breast cancer stem cells. Cancer Res. 69, 2887–2895.

Sark, M.W., Fischer, D.F., de Meijer, E., et al., 1998. AP-1 and Ets transcription factors regulate the expression of the human SPRR1A keratinocyte terminal differentiation marker. J. Biol. Chem. 273, 24683–24692.

Sato, S., Noguchi, Y., Ohara, N., et al., 2007. Identification of XAGE-1 isoforms: predominant expression of XAGE-1b in testis and tumors (11 authors). Cancer Immun. 7, 5.

Sauter, M., Schommer, S., Kremmer, E., et al., 1995. Human endogenous retrovirus K10: expression of gag protein and detection of antibodies in patients with seminomas. J. Virol. 69, 414–421.

Sayers, E.W., Barrett, T., Benson, D.A., et al., 2012. Database resources of the National Center for Biotechnology Information (40 authors). Nucl. Acids Res. 40, D13–D25.

Saze, H., 2008. Epigenetic memory transmission through mitosis and meiosis in plants. Semin. Cell Dev. Biol. 19, 527–536.

Scanlan, M.J., Gordon, C.M., Williamson, B., et al., 2002. Identification of cancer/testis genes by database mining and mRNA expression analysis. Int. J. Cancer 98, 485–492.

Scanlan, M.J., Gure, A.O., Jungbluth, A.A., et al., 2002b. Cancer/testis antigens: an expanding family of targets for cancer immunotherapy. Immunol. Rev. 188, 22–32.

Scanlan, M.J., Simpson, A.J., Old, L.J., 2004. The cancer/testis genes: review, standardization and commentary. Cancer Immun. 4, 1–15.

Schafer, M., Werner, S., 2008. Cancer as an overhealing wound: an old hypothesis revisited. Nat. Rev. Mol. Cell Biol. 9, 628–638.

Schauenstein, G., 1908. Histologische Untersuchungen uber atypisches Plattenepithel an der Portio und an der Innenflache der Cervix. Arch. J. Gynak. 85, 576–616.

Schier, A.F., 2003. Nodal signaling in vertebrate development. Annu. Rev. Cell Dev. Biol. 19, 589–621.

Schierwater, B., Kuhn, K., 1998. Homology of Hox genes and the zootype concept in early metazoan evolution. Mol. Phyl. Evol. 9, 375–381.

Schlom J., Palena, C.M., Kozlov, A.P., et al., 2012. Brachyury polipeptides and methods for use. United States Patent No. 8,188,214 B2.

Schlumberger, H.G., 1952. Nerve sheath tumors in an isolated goldfish population. Cancer Res. 12, 890–899.

Schmale, H., Bamberger, C., 1997. A novel protein with strong homology to the tumor suppressor p53. Oncogene 15, 1363–1367.

Schmid, K.J., Aquadro, C.F., 2001. The evolutionary analysis of "orphans" from *Drosophila* genome identifies rapidly diverging and incorrectly annotated genes. Genetics 159, 589–598.

Schmidt, E.E., 1996. Transcriptional promiscuity in testes. Curr. Biol. 6, 768–769.

Schmitz, R.J., Schultz, M.D., Lewsey, M.G., et al., 2011. Transgenerational epigenetic instability is a source of novel methylation variants. Science 334, 369–373.

Schneider, S.Q., Finnerty, J.R., Martindale, M.Q., 2003. Protein evolution: structure-function relationship of the oncogene beta-catenin in the evolution of multicellular animals. J. Exp. Zool. (Mol. Dev. Evol.) 295B, 25–44.

Schoen, E.J., DiRainondo, V.D., Dominguez, O.V., 1961. Bilaterial testicular tumors complicating adrenocortical hyperplasia. J. Clin. Endocr. 21, 518–522.

Schreiber, R.D., Old, L.J., Smyth, M.J., 2011. Cancer immunoediting: integrating immunity's roles in cancer suppression and promotion. Science 331, 1565–1570.

Schubeler, D., 2012. Epigenetic islands in a genetic ocean. Science 338, 756–757.

Schubert, D., Humphreys, S., Baroni, C., et al., 1969. *In vitro* differentiation of a mouse neuroblastoma. Proc. Nat. Acad. Sci. USA 64, 316–323.

Schubert, D., Humphreys, S., de Vitry, F., et al., 1971. Induced differentiation of a neuroblastoma. Dev. Biol. 25, 514–546.

Schubert, F.R., Nieselt-Struwe, K., Gruss, P., 1993. The Antennapedia-type homeobox genes have evolved from three precursors separated early in metazoan evolution. Proc. Natl. Acad. Sci. USA 90, 143–147.

Schulz, W.A., 1998. DNA methylation in urological malignancies. Int. J. Oncol. 13, 151–167.

Schulz, W.A., 2006. L1 retrotransposons in human cancers. J. Biomed. Biotechnol. 2006, 83672.

Schulz, W.A., Hoffmann, M.J., 2009. Epigenetic mechanisms in the biology of prostate cancer. Sem. Cancer Biol. 19, 172–180.

Scubbert, S., Shannon, K., Bollag, G., 2007. Hyperactive Ras in developmental disorders and cancer. Nat. Rev. Cancer 7, 295–308.

Sela, N., Mersch, B., Gal-Mark, N., et al., 2007. Comparative analysis of transposed elements' insertion within human and mouse genomes reveals *Alu*'s unique role in shaping the human transcriptome. Genome Biol. 8. Available from: http://dx.doi.org/10.1186/gb-2007-8-6-r127.

Selbach, M., Schwanhäusser, B., Thierfelder, N., et al., 2008. Widespread changes in protein synthesis induced by microRNAs. Nature 455, 58–63.

Sell, C., 2003. Caloric restriction and insulin-like growth factors in aging and cancer. Horm. Metab. Res. 35, 705–711.

Sell, S., 2004. Stem cell origin of cancer and differentiation therapy. Critical Rev. Oncol. Hematol. 51, 1–28.

Sell, S., Leffert, H.L., 2008. Liver cancer stem cells. J. Clin. Oncol. 26, 2800–2805.

Severtsov, A.N., 1927. Über die Beziehungen zwischen der Ontogenese und der Phylogenese der Tiere, Jena. Z. Naturwiss. 56, 51–180 (o.s. 63).

Severtsov, A.N., 1935. Modusy Filembriogeneza, Zool. Zhur 14, 1—8.

Severtsov, A.N., 1949. Morphological Laws of Evolution. Collection of Works, vol. 5. Publishing house of the Academy of sciences of the U.S.S.R., Moscow, Leningrad.

Shabad, L.M., 1967. Experimental and Morphological Aspects of Precancerous Lesions. Medicine, Moscow, 384 p.

Shakya, R., Reid, L.J., Reczek, C.R., et al., 2011. BRCA1 tumor suppression depends on BRCT phosphoprotein binding, but not its E3 ligase activity. Science 334, 525—528.

She, X., Cheng, Z., Zollner, S., et al., 2008. Mouse segmental duplication and copy number variation. Nat. Genet. 40, 909—914.

Shilo, B.-Z., Weinberg, R.A., 1981. DNA sequences homologous to vertebrate oncogenes are conserved in *Drosophila melanogaster*. Proc. Natl. Acad. Sci. USA 78, 6789—6792.

Shilov, E.S., Murashev, B.V., Popovich, I.G., et al., 2009. The perspective novel model of fish tumors. Russ. J. AIDS, Cancer and Public Health 13, 7, http://www.aidsconference.spb.ru/articles/arc9UuPPc.pdf.

Shimeld, S.M., Holland, P.W.H., 2000. Vertebrate innovations. Proc. Natl. Acad. Sci. USA 97, 4449—4452.

Shinkaruk, S., Bayle, M., Lain, G., et al., 2003. Vascular endothelial growth factor (VEGF), an emerging target for cancer chemotherapy. Curr. Med. Chem. Anticancer Agents 3, 95—117.

Shishegar, M., Ashraf, M.J., Azarpira, N., et al., 2011. Salivary gland tumors in maxillofacial region: a retrospective study of 130 cases in a Southern Iranian population. Path. Res. Interntl. Available from: http://dx.doi.org/10.4061/2011/934350.

Shubin, N., Tabin, C., Carroll, S., 2009. Deep homology and the origins of evolutionary novelty. Nature 457, 818—823.

Shvemberger, I.N., 1986. Conversion of malignant cells into normal ones. Int. Rev. Cytol. 103, 341—386.

Siedlecki, M., 1901. Sur les rapports des gregarines avec l'epithelium intestinal. CR Soc. Biol. (Paris) 53, 81—83.

Sikandar, S.S., Pate, K.T., Anderson, S., et al., 2010. *NOTCH* signaling is required for formation and self-renewal of tumor-initiating cells and for repression of secretory cell differentiation in colon cancer. Cancer Res. 70, 1469—1478.

Sikaroodi, M., Galachiantz, Y., Baranova, A., 2010. Tumor markers: the potential of "Omics" approach. Curr. Mol. Med. 10, 249—257.

Simpson, A.J., Caballero, O.L., Jungbluth, A., et al., 2005. Cancer/testis antigens, gametogenesis and cancer. Nat. Rev. Cancer 5, 615—625.

Sinderman, C.J., 1990. Neoplastic diseases. second ed. In: Sinderman, C.J. (Ed.), Principal Diseases of Marine Fish and Shellfish, vol. 1. Academic Press, San Diego, pp. 173—199.

Singh, S.K., Clarke, I.D., Terasaki, M., et al., 2003. Identification of a cancer stem cell in human brain tumors. Cancer Res. 63, 5821—5828.

Skaletsky, H., Kuroda-Kawaguchi, T., Minx, P.J., et al., 2003. The male-specific region of the human Y chromosome is a mosaic of discrete sequence classes. Nature 423, 825—837.

Sly, B.J., Snoke, M.S., Raff, R.A., 2003. Who came first — larvae or adults? Origins of bilaterian metazoan larvae. Int. J. Dev. Biol. 47, 623—632.

Smit, A.F.A., 1996. The origin of interspersed repeats in the human genome. Curr. Opin. Genet. Dev. 6, 743—748.

Smith, E.F., Townsend, C.O., 1907. A plant-tumor of bacterial origin. Science 25, 671—673.

Smith, A., Ashworth, A., 1998. Cancer predisposition: where's the phosphate? Curr. Biol. 8, R241—R243.

Smyth, G.E., Stern, K., 1938. Tumors of the thalamus: a clinicopathological study. Brain 61, 339–374.

Snel, B., Bork, P., Huinen, M.A., 2002. Genomes in flux: the evolution of Archaeal and Proteobacterial gene content. Genome Res. 12, 17–25.

Sorek, R., 2007. The birth of new exons: mechanisms and evolutionary consequences. RNA 13, 1603–1608.

Sorek, R., Ast, G., Graur, D., 2002. *Alu*-containing exons are alternatively spliced. Genome Res. 12, 1060–1067.

Sparks, A.K., 1969. Review of tumors and tumor-like conditions in Protozoa, Coelenterata, Platyhelminthes, Annelida, Sipunculida, and Arthropoda, excluding Insects. Natl. Cancer Inst. Monogr. 31, 671–682.

Sparks, A.K., 1985. Synolpsis of Invertebrate Pathology Exclusive of Insects. Elsevier, Amsterdam.

Spiegelman, S., 1948. Differentiation as controlled production of unique enzymatic patterns. Symp. Soc. Exp. Biol. 2, 286–325.

Spiotto, M.T., Yu, P., Rowley, D.A., et al., 2002. Increasing tumor antigen expression overcomes "ignorance" to solid tumors via crosspresentation by bone marrow-derived stromal cells. Immunity 17, 737–747.

Srivastava, M., Simakov, O., Chapman, J., et al., 2010. The *Amphimedon queenslandica* genome and the evolution of animal complexity. Nature 466, 720–726.

Stange, D.E., Engel, F., Longerich, T., et al., 2010. Expression of ASCL2 related stem cell signature and IGF2 in colorectal cancer liver metastases with 11p15.5 gain. Gut 59, 1236–1244.

Stauffer, Y., Theiler, G., Sperisen, P., et al., 2004. Digital expression profiles of human endogenous retroviral families in normal and cancerous tissues. Cancer Immun. 4, 2.

Stavrodimos, K.G., Tyritzis, S.I., Migdalis, V., et al., 2010. Benign renal tumor prevalence and its correlation with patient characteristics and pathology report data. The Internet J. Urol 6 (2). Available from: http://dx.doi.org/10.5580/1754.

Steel, K., Baerg, R.D., Adams, D.O., 1967. Cushing's syndrome associated with carcinoid tumor of the lung. J. Clin. Endocr. 27, 1285–1289.

Steeves, H.R., 1969. An epithelial papilloma of the brown bullhead, *Ictalurus nebulosis*. Natl. Cancer Inst. Monogr. 31, 215–217.

Stevenson, B.J., Iseli, C., Panji, S., et al., 2007. Rapid evolution of cancer/testis genes on the X chromosome. BMC Genomics 8 (129). Available from: http://dx.doi.org/10.1186/1471-2164-8-129.

Stewart, H.L., 1941. Hyperplastic and neoplastic lesions of the stomach in mice. J. Nat. Cancer Inst. 1, 489–509.

Stoltz, D.B., Whitefield, J.B., 2009. Making nice with viruses. Science 323, 884–885.

Stranger, B.Z., Tanaka, A.J., Melton, D.A., 2007. Organ size is limited by the number of embryonic progenitor cells in the pancreas but not in the liver. Nature 445, 886–891.

Stransky, N., Egloff, A.M., Tward, A.D., et al., 2011. The mutational landscape of head and neck squamous cell carcinoma (39 authors). Science 333, 1157–1160.

Strathmann, R.R., 2000. Functional design in the evolution of embryos and larvae. Cell Dev. Biol. 11, 395–402.

Strauss, D.C., Thomas, M., 2010. Transmission of donor melanoma by organ transplantation. Lancet Oncol. 11, 790–796.

Subramanian, S., Kumar, S., 2004. Gene expression intensity shapes evolutionary rates of the proteins encoded by vertebrate genome. Genetics 168, 373–381.

Sulston, J.E., Schierenberg, E., White, J.G., et al., 1983. The embryonic cell lineage of the nematode *Caenorhabditis elegans*. Dev. Biol. 100, 64–119.

Summers, K., da Silva, J., Farwell, M.A., 2002. Intragenomic conflict and cancer. Med. Hypotheses 59, 170–179.

Supovit, S.C., Rosen, J.M., 1981. Tumor-specific polyadenylated RNAs from 7,12-dymethylbenz (a) anthracene-induced mammary tumors revealed through hybridization with fractionated single copy DNA. Cancer Res. 41, 3827–3834.

Syasin, I.G., Sokolovsky, A.S., Phedorova, M., 1999. Skin tumors in *Pleuronectes obscures* (*Pleuronectidae*) represent a complex combination of epidermal papilloma and rhabdomyosarcoma. Dis. Aquat. Organ. 39, 49–57.

Taft, R.J., Pheasant, M., Mattick, J.S., 2007. The relationship between non-protein-coding DNA and eukaryotic complexity. Bioessays 29, 288–299.

Taipale, J., Beachy, P.A., 2001. The Hedgehog and Wnt signaling pathways in cancer. Nature 411, 349–354.

Tam, O.H., Aravin, A.A., Stein, P., et al., 2008. Pseudogene-derived small interfering RNAs regulate gene expression in mouse oocytes. Nature 453, 534–538.

Tanaka, M., Levy, J., Terada, M., et al., 1975. Induction of erythroid differentiation in murine virus infected erythroleukemia cells by highly polar compounds. Proc. Natl. Acad. Sci. USA 72, 1003–1006.

Tatarinov, Y.S., 1964. Presence of embryo-specific alpha-globulin in serum of patients with primary hepatocellular carcinoma. Vopr. Med. Khim 10, 90–91.

Tautz, D., Domazet-Loso, T., 2011. The evolutionary origin of orphan genes. Nat. Rev. Genetics 12, 692–702.

Taylor, J.S., Raes, J., 2004. Duplication and divergence: the evolution of new genes and old ideas. Annu. Rev. Genet. 38, 615–643.

Tchakhotine, S., 1938. Cancerisation experimentale des elements embryonnaires obtenus sur les larves d'oursins. CR Soc. Biol. (Paris) 127, 1195–1197.

Thiery, J.P., 2003. Epithelial-mesenchymal transitions in development and pathologies. Curr. Opin. Cell Biol. 15, 740–746.

Thiery, J.P., Sleeman, J.P., 2006. Complex networks orchestrate epithelial-mesenchymal transitions. Nature Rev. Mol. Cell. Biol. 7, 131–142.

Thiery, J.P., Acloque, H., Huang, R.Y.J., et al., 2009. Epithelial-mesenchymal transitions in development and disease. Cell 139, 871–890.

Thomas, P., Toth, C.A., Saini, K.S., et al., 1990. The structure, metabolism and function of the carcinoembryonic antigen gene family. Biochim. Biophys. Acta. 1032, 177–189.

Thomas, M.A., Weston, B., Joseph, M., et al., 2003. Evolutionary dynamics of oncogenes and tumor suppressor genes: higher intensities of purifying selection than other genes. Mol. Biol. Evol. 20, 964–968.

Timpson, N., Heron, J., Smith, G.D., et al., 2007. Comment on papers by Evans *et al.*, and Mekel-Bobrov *et al.* on evidence for positive selection of *MCPH1* and *ASPM*. Science 317, 1036a.

Toll-Riera, M., Bosch, N., Bellora, N., et al., 2009. Origin of primate orphan genes: a comparative genomics approach. Mol. Biol. Evol. 26, 603–612.

Topczewska, J.M., Postovit, L.-M., Margaryan, N.V., et al., 2006. Embryonic and tumorigenic pathways converge via Nodal signaling: role in melanoma aggressiveness. Nat. Med. 12, 925–932.

Torosian, M.H., 1988. The clinical usefulness and limitations of tumor markers. Surg. Gyn. Obst. 166, 567–579.

Torrents, D., Suyama, M., Zdobnov, E., et al., 2003. A genome-wide survey of human pseudogenes. Genome Res. 13, 2559–2567.

Toshida, H., Mamada, N., Fujimaki, T., et al., 2012. Incidence of benign and malignant eyelid tumors in japan. Int. J. Ophthalmic Pathol. 1, 2. Available from: http://dx.doi.org/10.4172/2324-8599.1000102.

Townson, J.L., Chambers, A.F., 2006. Dormancy of solitary metastatic cells. Cell Cycle 5, 1744–1750.

Turnbaugh, P.J., Ley, R.E., Hamady, M., et al., 2007. The human microbiome project. Nature 449, 804–810.

Twardzik, D.R., Ranchalis, J.E., McPherson, J.M., et al., 1989. Inhibition and promotion of differentiated-like phenotype of a human lung carcinoma in athymic mice by natural and recombinant forms of transforming growth factor-beta. J. Natl. Cancer Inst. 81, 1182–1185.

Valentine, J.W., 1994. Late Precambrian bilaterians: grades and clades. Proc. Natl. Acad. Sci. USA 91, 6751–6757.

Valentine, J.W., Collins, A.G., 2000. The significance of moulting in Ecdysozoan evolution. Evol. Dev. 2, 152–156.

van Beneden, R.J., 1997. Activated oncogenes in fish and molluscan neoplasms. In: Rossi, L., Richardson, R., Harshberger, J. (Eds.), Spontaneous Animal Tumors: A Survey. Press Point di Abbiategrasso, Milano, pp. 65–71.

van Kuilenburg, A.B.P., Dobritzsch, D., Meijer, J., et al., 2010. Dihydropyrimidinase deficiency: phenotype, genotype and structural consequences in 17 patients (22 authors). Biochim. Biophys. Acta. 1802, 639–648.

van Rijk, A., Bloemendal, H., 2003. Molecular mechanisms of exon shuffling: illegitimate recombination. Genetica 118, 245–249.

van Rijk, A.A., de Jong, W.W., Bloemendal, H., 1999. Exon shuffling mimicked in cell culture. Proc. Natl. Acad. Sci. USA 96, 8074–8079.

Van de Peer, Y., Maere, S., Meyer, A., 2009. The evolutionary significance of ancient genome duplications. Nat. Rev. Genet. 10, 725–732.

van der Bruggen, P., Traversari, C., Chomez, P., et al., 1991. A gene encoding an antigen recognized by cytolytic T lymphocytes on a human melanoma. Science 254, 1643–1647.

Venter, J.C., Adams, M.D., Myers, E.W., et al., 2001. The sequence of human genome. Science 291, 1304–1351.

Vermeulen, L., Todaro, M., de Sousa Mello, F., et al., 2008. Single-cell cloning of colon cancer stem cells reveals a multi-lineage differentiation capacity. Proc. Natl. Acad. Sci. USA 105, 13427–13432.

Vick, N.A., Lin, M.J., Bigner, D.D., 1977. The role of the subependymal plate in glial tumorigenesis. Acta Neuropathol. 40, 63–71.

Vickaryous, M.K., Hall, B.K., 2006. Human cell type diversity, evolution, development, and classification with special reference to cells derived from neural crest. Biol. Rev. Camb. Philos. Soc. 81, 425–455.

Vicovac, L., Aplin, J.D., 1996. Epithelial-mesenchymal transition during trophoblast differentiation. Acta Anat. 156, 202–216.

Vincent, M.D., 1985. The clinical problem. In: Farmer, P.B., Walker, J.M. (Eds.), The Molecular Basis of Cancer. Croom Helm, London and Sidney, pp. 1–35.

Vinckenbosch, N., Dupanloup, I., Kaessmann, H., 2006. Evolutionary fate of retroposed gene copies in the human genome. Proc. Natl. Acad. Sci. USA 103, 3220–3225.

Virchow, R., 1860. Cellular Pathology. John Churchill, New Burlington Street, London.

Visvader, J.E., Lindeman, G.J., 2008. Cancer stem cells in solid tumors: accumulating evidence and unresolved questions. Nat. Rev. Cancer 8, 755–768.

von Baer, K.E., 1828. Entwicklungsgeschichte der Thiere: Beobachtung und Re-flexion. Bornträger, Königsberg, 264 pp.

von Both, I., Silvestri, C., Erdemir, T., et al., 2004. Foxh1 is essential for development of the anterior heart field. Dev. Cell 7, 331–345.

Vogel, G., 2011. Do jumping genes spawn diversity? Science 332, 300–301.

Vogel, C., Chothia, C., 2006. Protein family expansions and biological complexity. PLoS Comp. Biol. 2, e48.

Voight, B.F., Kudaravalli, S., Wen, X., et al., 2006. A map of recent positive selection in the human genome. PLoS Biol. 4 (e72), 0446–0458.

Volff, J.-N., 2006. Turning junk into gold: domestication of transposable elements and the creation of new genes in eukaryotes. BioEssays 28, 913–922.

Vorontsov, N.N., 2003. Macromutations and evolution: fixation of Goldschmidt's macromutations as species and genus characters. Papillomatosis and appearance of macrovilli in the rodent stomach. Russ. J. Genet. 39, 422–426.

Waddington, C.H., 1940. Organizers and Genes. Cambridge Univ. Press, Cambridge.

Waddington, C.H., 1948. The genetic control of development. Symp. Soc. Exp. Biol. 2, 145–154.

Waddington, C.H., 1966. Fields and gradients. In: Locke, M. (Ed.), Major Problems in Developmental Biology. Academic Press, New York, pp. 105–124.

Wagner, A., 2000. Decoupled evolution of coding region and mRNA expression patterns after gene duplication: implications for the neutralist-selectionist debate. Proc. Natl. Acad. Sci. USA 97, 6579–6584.

Wagner, A., 2001. The yeast protein interaction network evolves rapidly and contains few redundant duplicate genes. Mol. Biol. Evol. 18, 1283–1292.

Wagner, A., 2005. Energy constraints on the evolution of gene expression. Mol. Biol. Evol. 22, 1365–1374.

Wagner, A., 2008. Gene duplication, robustness and evolutionary innovations. BioEssays 30, 367–373.

Wagner, D.E., Wang, I.E., Reddien, P.W., 2011. Clonogenic neoblasts are pluripotent adult stem cells that underlie planarian regeneration. Science 332, 811–816.

Walker, R., 1969. Virus associated with epidermal hyperplasia in fish. Natl. Cancer Inst. Monogr. 31, 195–207.

Walsh, J.B., 1995. How often do duplicated genes evolve new functions? Genetics 139, 421–428.

Walsh, J.B., Stephan, W., 2001. Multigene Families: Evolution. Encyclopedia of Life Sciences, Nature Publishing Group, www.els.net.

Wang, Z.Y., Chen, Z., 2008. Acute promyelocytic leukemia: from highly fatal to highly curable. Blood 111, 2505–2515.

Wang, W., Brunet, F.G., Nevo, E., et al., 2002. Origin of sphinx, a young chimeric RNA gene in Drosophila melanogaster. Proc. Natl. Acad. Sci. USA 99, 4448–4453.

Wang, W., Zheng, H., Yang, S., et al., 2005. Origin and evolution of new exons in rodents. Genome Res. 15, 1258–1264.

Wang, E.T., Kodama, G., Baldi, P., et al., 2006. Global landscape of recent inferred Darwinian selection for Homo sapiens. Proc. Natl. Acad. Sci. USA 103, 135–140.

Wang, H.Y., Chien, H.C., Osada, N., et al., 2007. Rate of evolution in brain-expressed genes in humans and other primates. Plos Biol. 5, e13.

Wang-Johanning, F., Frost, A.R., Johanning, G.L., et al., 2001. Expression of human endogenous retrovirus k envelope transcripts in human breast cancer. Clin. Cancer Res. 7, 1553–1560.

Wang-Johanning, F., Frost, A.R., Jian, B., et al., 2003. Detecting the expression of human endogenous retrovirus E envelope transcripts in human prostate adenocarcinoma. Cancer 98, 187–197.

Wapinsky, I., Pfeffer, A., Friedman, N., et al., 2007. Natural history and evolutionary principles of gene duplication in fungi. Nature 449, 54–61.

Warburton, P.E., Giodano, J., Cheung, F., et al., 2004. Inverted repeat structure of the human genome: the X-chromosome contains a preponderance of large, highly homologous inverted repeats that contain testes genes. Genome Res. 14, 1861–1869.

Watanabe, T., Totoki, Y., Toyoda, A., et al., 2008. Endogenous siRNAs from naturally formed dsRNAs regulate transcripts in mouse oocytes. Nature 453, 539–543.

Waterhouse, R.M., Zdobnov, E.M., Kriventseva, E.V., 2010. Correlating traits of gene retention, sequence divergence, duplicability and essentiality in vertebrates, arthropods and fungi. Genome Biol. Evol. 2, 75–86.

Waxman, S., Rossi, G.B., Takaku, F., 1988. The Status of Differentiation Therapy of Cancer. Raven, New York.

Weaver, V.M., Petersen, O.W., Wang, F., et al., 1997. Reversion of malignant phenotype of human breast cells in three-dimensional culture and *in vivo* by integrin blocking antibodies. J. Cell Biol. 137, 231–245.

Weinberg, R.A., 1985. The action of oncogenes in the cytoplasm and nucleus. Science 230, 770–776.

Weis, S., Schartl, M., 1998. The macromelanophore locus and the melanoma oncogene Xmrk are separate genetic entities in the genome of Xiphophorus. Genetics 149, 1909–1920.

Weismann, A., 1893. The Germ-Plasm: A Theory of Heredity. Charles Scribner's Sons, New York.

Welch, H.G., Black, W.C., 1997. Using autopsy series to estimate the disease "reservoir" for ductal carcinoma *in situ* of the breast: how much more breast cancer can we find? Ann. Intern. Med. 127, 1023–1028.

Wellings, S.R., 1969. Neoplasia and primitive vertebrate phylogeny: echinoderms, prevertebrates, and fishes – a review. Natl. Cancer Inst. Monogr. 31, 59–128.

Wells, H.G., 1940. Occurrence and significance of congenital malignant neoplasms. Arch. Path. 30, 535–601.

Whitesell, L., Mimnaugh, E.G., De Costa, B., et al., 1994. Inhibition of heat shock protein HSP90-pp60[v-src] heteroprotein complex formation by benzoquinone ansamycins: essential role for stress proteins in oncogenic transformation. Proc. Natl. Acad. Sci. USA 91, 8324–8328.

Wiemann, B., Starnes, C.O., 1994. Coley's toxins, tumor necrosis factor and cancer research: a historical perspective. Pharmacol. Ther. 64, 529–564.

Wikipedia, 2010. http://en.wikipedia.org/wiki/Evolution_of_mammalian_auditory_ossicles.

Wildgoose, W.H., 1992. Papilloma and squamous cell carcinoma in koi carp (*Cyprinus carpio*). Vet. Rec. 130, 153–157.

Wildman, D.E., Chen, C., Erez, O., et al., 2006. Evolution of the mammalian placenta revealed by phylogenetic analysis. Proc. Natl. Acad. Sci. USA 103, 3203–3208.

Williams, J.W., 1890. A tumor in the fresh-water mussel, *Anodonta cygnea* Linn. J. Anat. Physiol. Norm. Path. 24, 307–308.

Williams, G.C., Nesse, R.M., 1991. The dawn of Darwinian medicine. Q. Rev. Biol. 66, 1–22.

Williams, S.E., Beronja, S., Pasolli, H.A., et al., 2011. Asymmetric cell divisions promote Notch-dependent epidermal differentiation. Nature 470, 353–358.

Williamson, S.H., Hubisz, M.J., Ckark, A.G., et al., 2007. Localizing recent adaptive evolution in the human genome. PLoS Genet. 3 (e90), 0901–0915.

Willis, R.A., 1953. Pathology of Tumors, second ed. Butterworth, London.

Willis, R.A., 1967. Pathology of Tumors, fourth ed. Butterworth, London.

Willmer, E.N., 1970. Cytology and Evolution. Academic Press, New York and London.

Woese, C.R., 1967. The Genetic Code. The Molecular Basis for Genetic Expression. Harper and Row, Publishers, New York, Evanston, London.

Wolfe, K.H., Li, W.H., 2003. Molecular evolution meets the genomics revolution. Nat. Genet. Suppl. 33, 255–265.

Wolpert, L., 1999. From egg to adult to Larva. Evol. Develop. 1, 3–4.

Woods, C.G., Bond, J., Enard, W., 2005. Autosomal recessive primary microcephaly (MCPH): a review of clinical, molecular, and evolutionary findings. Am. J. Hum. Genet. 76, 717–728.

Wu, X., Ruvkun, G., 2010. Germ cell genes and cancer. Science 330, 1761–1762.

Wu, D.D., Irwin, D.M., Zhang, Y.P., 2011. De novo origin of human protein-coding genes. PLoS Genet. 7, e1002379.

Wu, X., Northcott, P.A., Dubuc, A., et al., 2012a. Clonal selection drives genetic divergence of metastatic medulloblastoma. Nature 482, 529–533.

Wu, Y.-C., Rasmussen, M.D., Kellis, M., 2012b. Evolution at subgene level: domain rearrangements in the *Drosophila* phylogeny. Mol. Biol. Evol. 29, 689–705.

Xiao, W., Liu, H., Li, Y., et al., 2009. A rice gene of *de novo* origin negatively regulates pathogen-induced defense response. PLoS ONE 4, e4603.

Xing, J., Wang, H., Belancio, V.P., et al., 2006. Emergence of primate genes by retrotransposon-mediated sequence transduction. Proc. Natl. Acad. Sci. USA 103, 17608–17613.

Xing, J., Zhang, Y., Han, K., et al., 2009. Mobile elements create structural variation: analysis of a complete human genome. Genome Res. 19, 1516–1526.

Xu, X., Lee, J., Stern, D.F., 2004. Microcephalin is a DNA damage response protein involved in regulation of CHK1 and BRCA1. J. Biol. Chem. 279, 34091–34094.

Yamamoto, K., Hayashi, Y., Hanada, R., et al., 1995. Mass screening and age-specific incidence of neuroblastoma in Saitama Prefecture, Japan. J. Clin. Oncol. 13, 2033–2038.

Yu, X., Chini, C.C.S., He, M., et al., 2003. The BRCT domain is a phospho-protein binding domain. Science 302, 639–642.

Yu, F.L., Hill, R.S., Schaffner, S.F., et al., 2007a. Comment on "Ongoing adaptive evolution of *ASPM*, a brain size determinant in *Homo sapience*." Science 316, 370b.

Yu, K., Liu, Y., Wang, H., et al., 2007b. Epidemiological and pathological characteristics of cardiac tumors: a clinical study of 242 cases. Interact. Cardio Vasc. Thorac. Surg. 6, 636–639.

Yuan, J., Horvitz, H.R., 1992. The *Caenorhabditis elegans* cell death gene *ced-4* encodes a novel protein and is expressed during the period of extensive programmed cell death. Development 116, 309–320.

Yuan, J., Shaham, S., Ledoux, S., et al., 1993. The *C. elegans* cell death gene *ced-3* encodes a protein similar to mammalian interleukin-1 beta-converting enzyme. Cell 75, 641–652.

Zaiss, D.M., Kloetzel, P.M., 1999. A second gene encoding the mouse proteasome activator PA28beta subunit is part of a LINE1 element and is driven by a LINE1 promoter. J. Mol. Biol. 287, 829–835.

Zavarzin, A.A., 1934. On the evolutionary dynamics of tissues. Arch. Biol. Sci. (U.S.S.R.) 36 (A), 3–64.

Zavarzin, A.A., 1953. Selected Works, vol. 4. Publishing House of the Academy of Sciences of the U.S.S.R., Moscow and Leningrad.

Zdanov, V.M., Tikhonenko, T.I., 1974. Viruses as a factor of evolution: exchange of genetic information in the biosphere. Adv. Virus Res. 19, 361–394.

Zecevic, N., Chen, Y., Filipovic, R., 2005. Contributions of cortical subventricular zone to the development of the human cerebral cortex. J. Comp. Neurol. 491, 109–122.

Zendman, A.J., Van Kraats, A.A., Weidle, U.H., et al., 2002. The XAGE family of cancer/testis-associated genes: alignment and expression in normal tissues, melanoma lesions and Ewing's sarcoma. Int. J. Cancer 99, 361–369.

Zendman, A.J., Zschocke, J., van Kraats, A.A., et al., 2003a. The human SPANX multigene family: genomic organization, alignment and expression in male germ cells and tumor cell lines. Gene 309, 125–133.

Zendman, A.J., Ruiter, D.J., Van Muijen, G.N., 2003b. Cancer/testis-associated genes: identification, expression profile, and putative function. J. Cell. Physiol. 194, 272–288.

Zerbini, L.F., Wang, Y., Czibere, A., 2004. NF-kB-mediated repression of growth arrest- and DNA-damage-inducible proteins 45α and γ is essential for cancer cell survival. Proc. Natl. Acad. Sci. USA 101, 13618–13623.

Zhang, J., 2003a. Evolution of the human ASPM gene, a major determinant of brain size. Genetics 165, 2063–2070.

Zhang, J., 2003b. Evolution by gene duplication: an update. Trends Ecol. Evol. 18, 292–298.

Zhang, J., 2006. Parallel adaptive origins of digestive RNases in Asian and African leaf monkeys. Nat. Genet. 38, 819–823.

Zhang, J., Rosenberg, H.F., 2002. Diversifying selection of the tumor-growth promoter angiogenin in primate evolution. Mol. Biol. Evol. 19, 438–445.

Zhang, J., Rosenberg, H.F., Nei, M., 1998. Positive Darwinian selection after gene duplication in primate ribonuclease genes. Proc. Natl. Acad. Sci. USA 95, 3708–3713.

Zhang, J., Zhang, Y., Rosenberg, H.F., 2002a. Adaptive evolution of a duplicated pancreatic ribonuclease gene in a leaf-eating monkey. Nat. Genet. 30, 411–415.

Zhang, J., Webb, D.M., Podlaha, O., 2002b. Accelerated protein evolution and origins of human-specific features: FOXP2 as an example. Genetics 162, 1825–1835.

Zhang, X., Sun, H., Danila, D.C., et al., 2002. Loss of expression of GADD45γ, a growth inhibitory gene, in human pituitary adenomas: implications for tumorigenesis. J. Clin. Endocrinol. Metab. 87, 1262–1267.

Zhang, X.H.-F., Chasin, L.A., 2006. Comparison of multiple vertebrate genomes reveals the birth and evolution of human exons. Proc. Natl. Acad. Sci. USA 103, 13427–13432.

Zhang, Z., Harrison, P.M., Liu, Y., et al., 2003. Millions of years of evolution preserved: a comprehensive catalog of the processed pseudogenes in the human genome. Genome Res. 13, 2541–2558.

Zhao, C., Blum, J., Chen, A., et al., 2007. Loss of β-catenin impairs the renewal of normal and CML stem cells in vivo. Cancer Cell 12, 528–541.

Zheng, D., Zhang, Z., Harrison, P.M., et al., 2005a. Integrated pseudogene annotation for human chromosome 22: evidence for transcription. J. Mol. Biol. 349, 27–45.

Zheng, P.Z., Wang, K.K., Zhang, Q.Y., et al., 2005b. Systems analysis of transcriptome and proteome in retinoic acid/arsenic trioxide-induced cell differentiation/apoptosis of promyelocytic leukemia. Proc. Natl. Acad. Sci. USA 102, 7653–7658.

Zhou, Q., Zhang, G., Xu, S., et al., 2008. On the origin of new genes in *Drosophila*. Genome Res. 18, 1446–1455.

Zhu, J., 2006. DNA methylation and hepatocellular carcinoma. J. Hepatobiliary Pancreat. Surg. 13, 265–273.

Ziegler, J.L., 1980. The occurrence of tumors in domestic animals, Editor in Chief. Natl. Cancer Inst. Monogr. 54, 210.

ZoomInfo, 2013. Dr. John C. Harshbarger. ZoomInfo.com.

Zupan, J., Muth, T.R., Draper, O., et al., 2000. The transfer of DNA from *Agrobacterium tumefaciens* into plants: a feast of fundamental insights. Plant J. 23, 11–28.

Index

Note: Page numbers followed by "*f*", and "*t*" refers to figures and tables respectively.

Printed and bound by CPI Group (UK) Ltd, Croydon, CR0 4YY

03/10/2024

01040420-0005